PROFESSOR R. W. FOGEL
HARVARD UNIVERSITY
1737 CAMBRIDGE ST., RM. G-7
CAMBRIDGE, MA. 02138

12/16/80

World Food Supply

This is a volume in the Arno Press collection

World Food Supply

Advisory Editor
D. Gale Johnson

Editorial Board
Charles M. Hardin
Kenneth H. Parsons

See last pages of this volume for a complete list of titles.

DEVELOPMENTS IN AMERICAN FARMING

ARNO PRESS
A New York Times Company
New York — 1976

Editorial Supervision: MARIE STARECK

Reprint Edition 1976 by Arno Press Inc.

Copyright © 1976 by Arno Press Inc.

Reprinted from copies in the State Historical
  Society of Wisconsin Library

WORLD FOOD SUPPLY
ISBN for complete set: 0-405-07766-1
See last pages of this volume for titles.

Manufactured in the United States of America

Library of Congress Cataloging in Publication Data
Main entry under title:

Developments in American farming.

   (World food supply)
   Reprint of the 1947 ed. of Progress of farm mechanization, by M. R. Cooper, G. T. Barton, and A. P. Brodell, published by U. S. Govt. Print. Off., Washington, as Miscellaneous publication no. 630 of the U. S. Dept. of Agriculture; and of the 1949 ed. of Changes in American farming, by S. E. Johnson, published by U. S. Govt. Print. Off., Washington, as Miscellaneous publication no. 707 of the U. S. Dept. of Agriculture.
   1. Agriculture--United States.   2. Farm mechanization --United States.   3. Agriculture--Economic aspects-- United States.   I. Cooper, Martin Reese, 1887- Progress of farm mechanization. 1976. II. Johnson, Sherman Ellsworth, 1896-   Changes in American farming. 1976. III. Series.   IV. Series: United States. Dept. of Agriculture. Miscellaneous publication ; no. 630. V. Series: United States. Dept. of Agriculture. Miscellaneous publication ; no. 707.
S441.D47   1976   338.1'6   75-29760
ISBN 0-405-07773-4

CONTENTS

Cooper, Martin R., Glen T. Barton and Albert P. Brodell
PROGRESS OF FARM MECHANIZATION (Miscellaneous Publication No. 630, United States Department of Agriculture), Washington, D. C., 1947

Johnson, Sherman E.
CHANGES IN AMERICAN FARMING (Miscellaneous Publication No. 707, United States Department of Agriculture), Washington, D. C., 1949

Progress of Farm Mechanization

By Martin R. Cooper,
Glen T. Barton, and Albert P. Brodell

Miscellaneous Publication No. 630
UNITED STATES DEPARTMENT OF AGRICULTURE
Washington 25, D. C., October 1947

FARM MECHANIZATION has made striking progress, and this progress, more than any other single thing, enabled American farmers to produce the large volume of agricultural products that helped so much to win World War II.

Some of the future potentialities of mechanization cannot be fully measured at this time. Others seem almost assured for the next farming generation.

This publication keynotes the place of farm mechanization in the greatest of all farm production jobs, and analyzes the influences of mechanization on farm employment, on efficiencies in production, and on production costs and returns of agriculture.

In all farm production, each farm worker in wartime in 1945 produced enough agricultural products to support himself and more than 13 others, whereas in 1920 one farm worker had supported himself and 9 other persons and in 1820, himself and only a little more than 3 other persons.

Each man-hour of farm labor meant 44 percent more gross production in 1945 than it did in 1917-21. Half of these savings in hours per unit of product resulted from mechanization. Other technological developments, primarily increases in yields of crops and livestock, were responsible for the other half.

Change in pattern of mechanization has been outstanding. Farm horses and mules have been rapidly replaced by tractors, trucks, and automobiles during the last third of a century. Combines, tractor-plows, tractor-cultivators, mechanical corn pickers, milking machines and other modern farming equipment are continuing to replace horse-drawn equipment and hand work. A modern tractor and its associated equipment now saves 850 hours of man labor compared with the time required with the animal power and equipment used a generation ago.

Thirty percent of the increase in food supplies for feeding an increasing population from 1920 to 1942 came from acreages released by the decline in horses and mules; 70 percent came from increased crop and livestock yields and from decreased exports. Crop production per acre has increased about one-fourth, and livestock production per unit of breeding stock has increased about one-third during the last quarter century. But crop acreages in 1944 were about the same as the 1917-21 average.

Mechanization has made possible more timely operations and this has contributed to increased farm production, especially in years when adverse weather delayed the preparation of land and planting.

The Corn Belt and the Great Plains are highly mechanized. Farms are generally less mechanized in the South than elsewhere. The next generation of farm people can expect some striking developments. Farm electrification and the use of new equipment proceeds at a rapid pace, and will aid farmers and farm women everywhere in doing things better and more easily.

Total physical production costs (labor, power, and other resources) per unit of farm output have decreased about 26 percent during the last quarter century and physical costs of labor, power, and machinery per unit of output about 30 percent. These large reductions in physical costs have benefited many individual farmers and agriculture as a whole.

The ratio of prices received by farmers to prices paid by farmers has fluctuated so violently that price changes have overshadowed increases in physical efficiency. During a large part of the period 1910–45 farm operators and family workers received, on an average, less net return per hour of labor than hired farm workers but many operators received additional agricultural income in the form of rent and interest on owned land and property.

Thus far, the march of farm technology has not always meant the release of farm workers. Labor-intensive enterprises and more production per acre and per animal have absorbed many of the people. This increased production is a principal cause of the increased physical efficiency in agriculture as a whole. Outlets for a large volume of farm products will be needed if farmers are to receive appropriate benefits from further increases in efficiency.

Progress of Farm Mechanization

By MARTIN R. COOPER, GLEN T. BARTON, and ALBERT P. BRODELL,
Agricultural Economists [1]

CONTENTS

	Page
Achievements of mechanization	2
Fewer farm workers needed	3
More output per farm worker	6
More product per hour of work	13
More food and fiber for human consumption	24
Less hand labor needed	28
Changes in pattern of mechanization	30
Displacement of work animals by mechanical power	33
Tractors and tractor equipment	36
Effect on timeliness of operation	41
Regional changes in mechanization	42
Growing importance of farm electrification	55
Mechanization and production costs and returns	56
The base period	56
Importance of power and machinery costs	58
Effect of prices on production costs	61
Agricultural costs and returns	69
Another 30 years of mechanization	75
Appendix	80

Peacetime prosperity and war activity always stimulate the demand for farm mechanization. The Civil War and the later settlement of the West accelerated the manufacture and improvement of labor-saving farm machines for preparing land, seeding, cultivating, and harvesting. By 1880, many farm machines—including hay presses and loaders and threshing machines—had important features of our modern machines, although refinements in design and construction were to follow. In World War I, with its race for food, and the large agricultural production that came in the 1920's sharp increases occurred on the farms in the number of the much-heralded tractors, motortrucks, and grain

[1] Members of the Bureau of Agricultural Economics who aided in assembling and appraising available data are R. W. Hecht, H. C. Norcross, Margaret F. Cannon, and J. W. Birkhead. Ada M. Procise was largely responsible for the computations.

combines and the popular hay loaders, manure spreaders, and labor-saving tillage, planting, and harvesting machines and tools. In World War II the urgent demand for farm machinery was only partially filled. But scarce as wartime farm machines were in relation to wartime needs, so many labor-saving machines were manufactured and sold that farmers now have more labor-saving machinery than at any other time in history.

Farm tractors that are as versatile as horses and mules in doing field and road jobs appeared on many more farms during the war, and some large productive farms added a second or third tractor. The number of grain combines, pick-up balers, side-delivery rakes, corn pickers, field forage harvesters, and milking machines actually increased in wartime, too. Now, new and better machines and tools are on the way to replace some of the finest machines and tools found anywhere in the world.

More work in less time and with less human effort—and on small farms as well as on large farms—is the central theme of modern farm mechanization. Less waste and better quality of product through better and more timely handling are motivating influences in the creation of better farm machines.

ACHIEVEMENTS OF MECHANIZATION

It has been said that the corn crop of 1840 was planted with the hoe, the plow, and the ax. This was probably literally true of those places where the land was being cleared of trees, or where the land was infested with stumps and roots. The hand-turned grindstone in the back yard that kept the farm tools sharp was a valued labor-saving item of farm mechanization.

As the cleared lands became free of obstacles, as the prairie sods were broken, and as improved machines came into use under rather favorable conditions, there was a gradual lessening of the hours of man labor required to produce a bushel of wheat or corn or a bale of cotton. Around 1880, our farmers produced a bushel of corn and a bale of cotton with little more than half the labor that had been used in 1800 and the time devoted to a bushel of wheat was cut to 40 percent. By 1940, about 190 hours were required to produce a bale of cotton, compared with the 300 hours some 60 years earlier. Even more progress had been made in regard to wheat and corn. Only 47 hours of direct man work were required to produce 100 bushels of wheat, compared with 152 hours in 1880; 83 hours were required to produce 100 bushels of corn, compared with 180 hours (table 1).

All of the decrease in the hours of work used in growing and harvesting a bushel of grain or a bale of cotton is not directly attributable to new or improved machines. Yields per acre have increased and this increase has called for proportionately less labor per bushel or pound. This is true especially of the crops that are harvested with machines. Expanded production in areas of low labor requirements has reduced the national average requirements per acre and per unit of production. Changes in cropping practices, even where the work is still done by hand, have frequently reduced the hours per unit of product. For example, a much larger part of the corn crop is harvested from the stand-

TABLE 1.—*Estimated man-hours used to produce an acre of wheat, corn, and cotton, and 100 bushels of wheat and corn, and a 500-pound gross weight bale of cotton for designated periods, United States average*

Crop and item	Yearly average for—					
	About 1800	About 1840	About 1880	About 1900	About 1920	About 1940
Wheat:						
Man-hours per acre before harvest	16	12	8	7	5.5	3.7
Harvest	40	23	12	8	6.5	3.8
Total	56	35	20	15	12.0	7.5
Yield per acre [1] ....bushels	15	15	13.2	13.9	13.8	15.9
Man-hours per 100 bushels	373	233	152	108	87	47
Corn for grain:						
Man-hours per acre before harvest	56	44	28	22	19	15
Harvest	30	25	18	16	13	10
Total	86	69	46	38	32	25
Yield per acre [1] ....bushels	25	25	25.6	25.9	28.4	30.3
Man-hours per 100 bushels	344	276	180	147	113	83
Cotton:						
Man-hours per acre before harvest	135	90	67	62	55	46
Harvest	50	45	52	50	35	52
Total	185	135	119	112	90	98
Yield of gross lint per acre [1] ....pounds	154	154	196	198	160	257
Man-hours per bale	601	439	304	283	281	191

[1] Yields for 1800 and 1840 are estimates by the authors. Yields for the other years are 5-year averages of published data, centered on year shown.

ing stalk by hand than in earlier days; formerly nearly all corn was cut, shocked, and then husked, and all by hand. In the first instance it takes only 6 or 7 hours to harvest an acre compared with 18 to 20 hours when the older methods were followed.

Fewer Farm Workers Needed

Ours was decidedly an agricultural nation for a long time after the colonial period. It has been estimated that nearly 72 percent of the working force of the United States in 1820 was engaged in agricultural pursuits, and only about 28 percent pursued nonagricultural livelihoods (table 2). As time passed the agricultural force represented a smaller and smaller proportion of the total labor force. By 1920, or shortly after the close of World War I, it was only 27 percent of the total force. Twenty years later, this percentage had dropped to 18. It is undoubtedly true, although we have no definite measure, that an even smaller percentage is now engaged in agriculture. In relation to the total population during the 120-year period from 1820 to 1940 the farm labor force found its numbers decreasing from about 22 percent to about 7 percent (table 2).

TABLE 2.—*Relation of labor force in agriculture to total labor force and total population of the United States, census years 1820–1940*

[Labor force includes persons 10 years old and over]

Year	Labor force [1]			Total population [2]	
	All occupations	Agricultural pursuits	Percentage farm labor force is of total	Total	Percentage farm labor force is of total population
	Number	Number	Percent	Number	Percent
1820	2,881,000	2,068,958	71.8	9,638,453	21.5
1830	3,931,537	2,772,453	70.5	12,866,020	21.5
1840	5,420,000	3,719,951	68.6	17,069,453	21.8
1850	7,697,196	4,901,882	63.7	23,191,876	21.1
1860	10,532,750	6,207,634	58.9	31,443,321	19.7
1870	12,924,951	6,849,772	53.0	38,558,371	17.8
1880	17,392,099	8,584,810	49.4	50,155,783	17.1
1890	23,318,183	9,938,373	42.6	62,947,714	15.8
1900	29,073,233	10,911,998	37.5	75,994,575	14.4
1910	37,370,794	11,591,767	31.0	91,972,226	12.6
1920	42,433,533	11,448,770	27.0	105,710,620	10.8
1930	48,829,920	10,471,998	21.4	122,775,046	8.5
1940	52,148,251	9,162,547	17.6	131,669,275	7.0

[1] Sixteenth Census of the United States: 1940 Series P-9, No. 11.
[2] Reports of the Bureau of the Census. Population data for 1820–1940 are from table 3, 1940 Census, Population, vol. II, part 1. During this period the month in which the census was taken varied, hence the month for which the population was reported varied.

It has been estimated that in 1820 each person on the farms in the United States produced enough food and fiber to support himself (or herself) and a little more than one-fourth of enough for an additional person (table 3). By the end of World War I, or by 1920, one farm person was producing enough to support himself, and two other persons, and nearly half enough for a fourth person. During the 20 years between the wars, 1920–40, production per farm person was again increased enough to support an additional one-half person, making about three extra persons. Then came a climactic increase during World War II. Each farm person produced enough to support, here and abroad, more than 5 and a half persons in 1945. In 5 years there was an increase per person of nearly 45 percent.

Persons supported can be measured in terms of people employed on farms as well as in terms of people living on farms. Measured in this way the increase per farm worker has been somewhat less than the increase in total persons supported per farm person but there have been variations from time to time (table 3). Until about 1910 a large percentage of the young people remained on the farms and immigrants of working age increased the farm-employment rolls. From 1910 to 1930 farm population and employment decreased at about the same rate. Depression increased this population. World War II broke out in Europe, and men and women, boys and girls began to leave the American farms by hundreds and flock to defense work. Then war broke on the United States and the hundreds became thousands. In all, 5 million people left the farms. But so large had been the farm population that total farm employment decreased by only 800,000 workers.

TABLE 3.—*Total farm population and farm employment, and average number of persons supported per farm person and per farm worker, United States, 1820–1945*

Year	Total farm population Jan. 1 [1]	Total farm employment [2]	Persons supported by one farm person [3]		Total persons supported at home and abroad by one farm worker [4]
			Supported persons of the United States	All persons supported at home and abroad	
	Millions	Millions	Number	Number	Number
1820	7.7	2.2	1.20	1.28	4.52
1830	9.8	2.9	1.27	1.35	4.51
1840	12.3	3.9	1.33	1.41	4.49
1850	15.8	5.1	1.44	1.51	4.68
1860	20.1	6.6	1.47	1.65	5.07
1870	22.4	7.2	1.60	1.78	5.60
1880	27.1	9.0	1.66	2.08	6.42
1890	29.4	10.4	1.84	2.26	6.59
1900	31.2	11.4	2.15	2.86	8.05
1910	32.1	12.1	2.54	2.97	7.99
1920	31.6	11.4	2.87	3.47	9.94
1930	30.2	11.2	3.61	4.01	10.96
1940	30.3	10.6	3.74	3.93	11.31
1945	25.2	9.8	5.08	5.64	14.54

[1] Data for 1910–45 taken from Bureau of Agricultural Economics report, Farm Population Estimates United States and Major Geographic Divisions, 1910–46, rounded to nearest 100 thousand. (Processed) June 1946. Data for 1820–1900 are estimates by the authors based largely on total population and numbers of persons engaged in agricultural pursuits.
[2] Data for 1910–45 taken from releases on farm employment issued by Bureau of Agricultural Economics, rounded to nearest 100 thousand. Data for 1820–1900 are estimates by the authors, based largely on the size of the labor force engaged in agricultural pursuits, table 2.
[3] Level of support at any given date is total food and fiber available for domestic consumption from farm production and from imports. The number of persons supported at home and abroad at the above level of support divided by the total farm population represents the total number of persons supported by one farm person.
[4] Total persons supported divided by total number of persons in farm employment in United States.

"Support" of these people by farmers has not meant the same thing throughout the 125 years for it is well known that in the early part of this period farm people did many things that were later done by city people. This was true of work in the farmhouse and on the farm.

But it is conceivable that the support furnished to consumers today may be greater than in early years when diets and clothing were simple if not meager. The changing composition of the diet has increased the volume of support supplied to consumers by farm people. Consumption of farm products per person increased, especially during World War II and immediately thereafter.

Perhaps achievements can best be measured by scanning further the century of mechanization that has now passed into history.

According to one enthusiastic writer, by 1840 improved plows, harrows, and other implements, and the introduction of rollers, cultivators, and drill-barrows and many additional items had made possible the working of a farm with half the labor that had been generally used 40 years before. But somewhat later, another writer said that an Ohio farmer of 1860 who used the best methods of production then available could probably produce his crops with two-thirds the labor required in 1840.

The year 1840 marks the beginning of worth-while results by inventors and experimenters who had been making persistent trials and studies throughout 50 years. The sickle of colonial days had virtually given way to modified types of the scythe and the grain cradle. In 1837 farmers were enthusiastic about the better cradles that were used in parts of the East, and the American scythe of that period, with its longer and thinner blade, was much used for several years after mowing machines were in the fields. Crude as the machines and tools of that time were compared with current models, they represented a long step forward from the beginning of the century.

Many makes of farm machines were tried out before 1850. Farm machines of the local blacksmith shop were being replaced by factory-made machines with wide commercial distribution. McCormick's reaper was becoming a reality. Portable horse-drawn power units and threshing machines were seen. The fanning mill, through with its critical period of experimentation, was in general use for cleaning grain. Horse-drawn mowing machines were at work in the hay fields. Shovel cultivators were replacing many steel hoes, and the steel walking plow was a few years old.

Then came the Civil War, the succeeding expansion, the Great War, the interwar period, and World War II with their described stimulations to mechanization.

More Output Per Farm Worker

The volume of farm output [2] in the United States by 1870 amounted to approximately 2.5 billion dollars. During the next 70 years, or by 1940, the volume of agricultural output had increased about 2.7 times, and amounted to 9.3 billion dollars in terms of 1870 agricultural price levels. This increase in agricultural output was accomplished with an increase of only 48 percent in farm employment (table 4). This large increase in agricultural output with so relatively small an increase in number of farm workers means that each worker increased his output 152 percent during the 70 years.

Several of the things that contributed to this huge increase in output per worker have been mentioned. Increased farm mechanization un-

---

[2] Farm output is gross farm production minus farm-produced power, and is approximately that part of total production which is available for human use. Farm-produced power is the cost in constant dollars of raising and maintaining farm horses and mules. For a full discussion of farm output, gross farm production, and farm-produced power, see Farm Production in War and Peace. By Glen T. Barton and Martin R. Cooper. Bur. of Agr. Econ. F. M. 53. (Processed).

TABLE 4—*Indexes of farm output, volume of farm power, machinery and equipment, farm employment, and farm population, United States, census years 1870–1940 and 1944–46*

[1870 = 100]

Year	Indexes of volume of farm output		Indexes of volume of farm power, machinery, and equipment		Farm output per unit of farm power, machinery and equipment [5]	Index of farm employment [6]	Index of farm population [7]
	Total [1]	Per worker [2]	Total [3]	Per worker [4]			
1870	100	100	100	100	100	100	100
1880	156	125	163	130	96	125	121
1890	183	125	232	159	79	146	131
1900	240	151	295	186	81	159	139
1910	261	154	385	228	68	169	143
1920	313	197	477	300	66	159	141
1930	324	208	471	302	69	156	135
1940	373	252	434	293	86	148	135
1944	446	319	494	353	90	140	114
1945	440	321	517	377	85	137	112
1946	453	324	542	387	84	140	116

[1] 1870–1910 based on "Ideal index," see table 61, p. 126 of U. S. Dept. Agr. Technical Bulletin 703, Gross Farm Income and Indices of Farm Production and Prices in the United States, 1869–1937. By Frederick Strauss and Louis H. Bean. 1940.
1920–44, based on index of "Farm Output" contained in Bureau of Agricultural Economics Farm Management Report 53. Glen T. Barton and Martin R. Cooper. op. cit.
[2] Index of total farm output divided by index of farm employment.
[3] Volume is in terms of 1935–39 average farm prices of all horses and mules (including harness) on farms 1870–1946, plus values of all farm machinery and equipment in terms of the 1935–39 wholesale price index, 1870–1910, and in terms of 1935–39 average farm prices 1911–46.
[4] Index of total volume of farm power, machinery, and equipment, divided by index of farm employment.
[5] Index of farm output divided by index of total volume of farm power, machinery, and equipment.
[6] See footnote 2, table 3.
[7] See footnote 1, table 3.

doubtedly contributed a lot directly to output per worker, and has made possible much of the Nation's rapid progress in expanding various lines of production that are well suited to mechanical production. An indication of mechanical possibilities in these directions is shown by the increasing volume of power and machinery on farms (table 4). In 1940, for example, the total volume of farm power (animal and mechanical), farm machinery, and farm equipment was 334 percent larger than the volume in 1870. As the number of workers engaged in agriculture increased during the 70 years by 48 percent, the volume of farm power, machinery, and equipment per worker increased less rapidly than the total increase, or about 193 percent.

In the war years, 1940–45, farm employment decreased 11 index points, and total farm power and machinery increased 83 index points. These changes resulted in an increase in volume of machinery and power per worker of 84 index points, or 30 percent. During the same period total agricultural output increased about 18 percent. During 1946, the first full year after the war, farm employment increased about 2 percent; the total volume of power, machinery, and equipment continued to increase; and the volume of farm output was larger than in any year during or before the war.

TABLE 5.—*Farm employment, man-hour requirements, and gross farm production per worker, United States, selected periods and years, 1909 to 1945*

[For indexes, 1917–21 = 100]

Period or year	Average annual farm employment		Man-hour requirements		Man-hours per worker		Gross production per worker
	Number of workers [1]	Index	Number [2]	Index	Number	Index	Index
	Thousands		*Mil. hrs.*		*Hours*		
1909–13	12,094	106	22,262	97	1,840	91	88
1917–21	11,403	100	22,983	100	2,016	100	100
1927–31	11,233	99	22,193	97	1,976	98	108
1939	10,740	94	20,454	89	1,904	94	118
1940	10,585	93	20,412	89	1,928	96	122
1941	10,361	91	20,617	90	1,990	99	129
1942	10,397	91	21,132	92	2,032	101	142
1943	10,263	90	21,026	91	2,049	102	140
1944	10,037	88	21,182	92	2,110	105	149
1945	9,844	86	20,655	90	2,098	104	152
Percentage change:							
1917–21 to 1939	−6	....	−11	....	−6	....	18
1939 to 1944	−6	....	3	....	12	....	26
1939 to 1945	−9	....	1	....	11	....	29

[1] Farm employment as estimated by the Bureau of Agricultural Economics. Workers (farm operators, unpaid members of their families, and hired workers) doing 2 or more days of farm work during the week of inquiry each month are counted in farm employment.

[2] Estimated man-hour requirements are in terms of time required by average adult males to do various farm tasks. As many women and children do less work in an hour than an average adult male, actual hours of work required are in excess of those shown in this table. The estimates of man-hour requirements for the years before 1939 are based largely on data contained in the WPA National Research Project report, "Changing Technology and Employment in Agriculture," by John A. Hopkins. Since this report was prepared some slight revisions have been made in the estimates of total man-hour requirements for 1939–45, and revisions are being made of the requirements shown for periods previous to 1939. The unrevised series has been used throughout this publication.

Modern wars seriously disturb the farm labor force. Women and youths are called upon to do more of the work in the fields and barns, and older men to work longer hours—and faster—as the younger men are drawn into the armed forces and into factories. The great war that ended in 1945 was no exception. In each year of the conflict farmers and their families worked hard to overcome the scarcity of labor. More days of work and more hours a day, fuller use of domestic and imported seasonal workers, and the slighting and elimination of some farm tasks, enabled fewer workers to produce more total food and fiber in each war year than we had ever produced in a year before. Good weather, good machines, good seeds, and other factors helped a lot, as will be shown later.

In 1945, each farm worker worked 11 percent more hours and turned out 29 percent more gross production than in 1939 (table 5). This average increase in gross production of 4.8 percent per year, with an average of 1.5 percent fewer workers, reflected a speeding up of trends that had been under way for many years. In the 20 years between the wars, 1917–21 to 1939, for example, gross production per worker increased 18 percent, or less than 1 percent per year compared with the 4.8 percent per year during World War II.

This achievement is all the more remarkable in view of the changing composition of the farm working force. The several million able-bodied men and experienced farm workers who left agriculture were not fully replaced in numbers even of less physically capable workers—older men, women, and children. The number of women farm workers almost doubled during wartime; thus, while total farm employment declined from 1939 to 1945 by about 9 percent, the average capacity of the farm labor force in terms of able-bodied men probably decreased considerably more.

From 1870 to 1920, the total volume of farm power and machinery combined increased more rapidly than the total volume of farm output [3]. In this 50-year period farm power, machinery, and equipment increased 177 percent more than farm output did, and in 1920 the index of farm output per unit of farm power, machinery, and equipment was only 66 compared with 100 in 1870 (table 4). From 1920 to 1946 the volume of agricultural output increased 45 percent but the volume of farm power, machinery, and equipment increased only about 14 percent. During most of the war and postwar years, 1940–46, the large increases in farm output were at least matched by increased volume of farm power, machinery, and equipment. This was achieved by large manufacture of tractors and other farm equipment in all years except 1943, and by the manufacture of a tremendous volume of repair parts for servicing old and new machines and equipment. In this way farmers managed to keep many tractors, automobiles, trucks, and other machines at work that ordinarily would have been junked.

Since 1910 there have been three important periods of farm mechanization. In that year mechanization had advanced to the point where farm employment began its long downward trend, even though total farm output continued its generally pronounced upward climb (fig. 1). The first period includes the years of World War I, and the first 10 or 12 years immediately following the war. The second period consists of the depression years of the 1930's, and the third period includes the years of the late 1930's and of World War II.

---

[3] It must be remembered that the measurement used here to show trend in volume of farm power, machinery, and equipment is in terms of constant prices, or in 1935–39 dollar values. This does not adequately measure the increase in horsepower with the coming in of tractors, trucks, and automobiles.

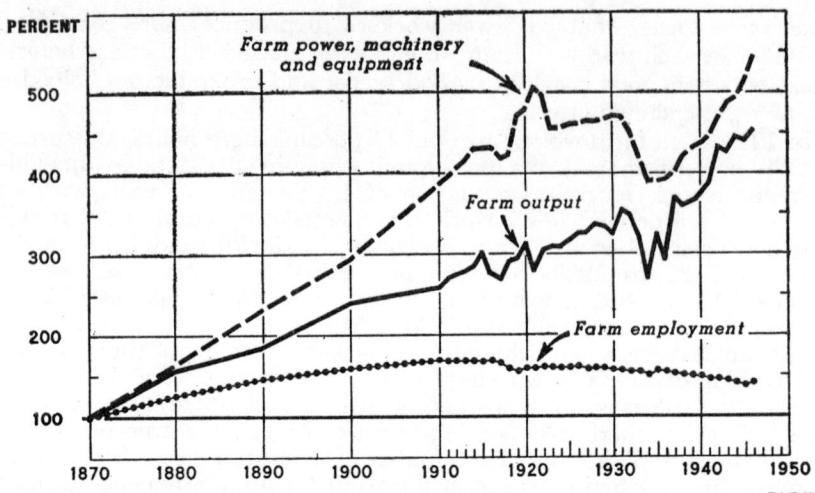

FIGURE 1—TOTAL VOLUME OF FARM POWER AND MACHINERY AND EQUIPMENT, FARM OUTPUT, AND TOTAL FARM EMPLOYMENT, UNITED STATES, 1870-1946. (VOLUME IN TERMS OF 1935-39 DOLLARS; INDEX NUMBERS, 1870=100)

Total volume of farm power—including horses and mules, machinery and equipment—has increased more rapidly than total farm output. The first high point of farm power, machinery, and equipment occurred in the early 1920's when farms were being mechanized and horse and mule numbers were being reduced somewhat. The recent wartime peak in volume resulted from large increases in numbers of tractors, trucks, and labor-saving machines.

The steep upward rise in farm mechanization that was so evident about 1910 was temporarily slowed down during World War I. But during the early conversion years after that war the steep rise was resumed. The peak was reached in 1921. This rise marked the beginning of the long and continuous transition from animal power and animal-drawn machines to tractor power and tractor equipment. The piling up in volume of farm power, machinery, and equipment just after World War I, therefore, was caused only in part by the continued upward surge in farm output, in relation to farm employment. As the transition progressed, and horses and mules and outmoded machines were gradually displaced by modern power units and machines, we needed less power, machinery, and equipment in terms of 1935-39 prices for a given volume of farm output.

With the industrial collapse of 1929 farmers reduced their machinery purchases. In the depression that followed, money to buy gasoline and expensive machines was very scarce; in places, horses and mules and horse-drawn machines were again used extensively, displacing tractors and power machines in the field. Restricted purchases of machines and equipment in the depression years were accompanied by reduced production. The severe droughts of 1934 and 1936 and some Government programs temporarily held down production of certain products.

But beginning in about 1937 the upward climb in farm output was resumed, and during World War II it was accelerated. Volume of power and machinery on farms turned upward also, and the rate of increase exceeded the rate of increase in farm output.

By 1942 there was only a 33-point spread between the indexes of farm output and farm power, machinery, and equipment, each index representing 100 percent in 1870. In 1920-21 the average spread had been 195 index points (fig. 1). Since 1942 the relatively large increase in farm power, machinery, and equipment has widened the spread somewhat. The closing up of the distance between the two indexes was caused by the change from animal power and equipment to mechanical power and equipment, in which greater work capacity per dollar of value in power and equipment was obtained, and by the larger production for human use from a given acreage of land.

Development of the general-purpose tractor and its complement of tools in the 1920's, and the extensive use of rubber tires in the late 1930's helped greatly in advancing the effectiveness of machines.

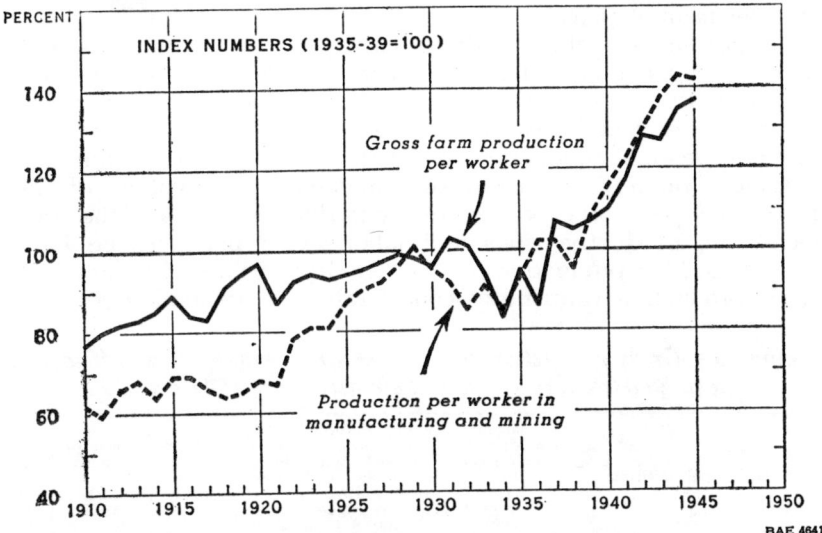

FIGURE 2—PRODUCTION PER WORKER IN AGRICULTURE AND INDUSTRY, UNITED STATES, 1910-45.

Over the long run, production per worker in manufacturing and mining has increased more than production per farm worker. But during the last 15 years increased production per worker in agriculture has been about the same as in manufacturing and mining.

This means that farmers were better supplied with labor-saving power units and machines in relation to production during World War II than is indicated by the indexes shown in figure 1. In the first place, a much larger percentage of the total volume of farm power and machinery was made up of modern labor-saving machines than was the case at any other time in our history. In the second place, increased production during this war was not the result of additional crop acres brought into cultivation, but came from increased crop yields and the feeding of more livestock. During the last years of the 1930's, and generally during the war many more of the labor-saving machines such as modern tractors and tractor equipment, combines, corn pickers, pick-up hay balers, and milking machines were at work on the farms, whereas horse-drawn machines decreased. More is said later about changes in numbers of tractors and other farm machines.

Indexes of gross farm production per worker or farm output plus farm-produced power, and of production per worker in manufacturing and mining, are available for each year from 1910 through 1945 (fig. 2). From 1910 to 1920, gross production per worker in agriculture increased more rapidly than production per worker in manufacturing and mining. From 1920 to 1929, the opposite was true; but since 1929 increase in farm production per worker has about kept pace with production per worker in manufacturing and mining. Even in the war years when industry was geared to overtime production, the increase per worker over the 1935–39 average was little more than the increased production per farm worker.

Longer hours in the fields and barns, help by the families, and relatively large supplies of labor-saving machines did the job. Because of better weather than the average, because more fertilizer and better seeds were used, and because large inventories of feeds from current and previous years of production were fed, relatively large production was obtained from a given amount of human effort. The volume of modern production per worker varies greatly in different parts of the country. Generally, productive soils and labor-saving machines go hand in hand with high production per worker. In some areas, volume of livestock production contributes considerably to production per worker.

TABLE 6.—*Gross production per worker and amount of specified capital items per worker, by geographic divisions, 1944 and 1945* [1]

Geographic division	Gross production per worker, 1944 (U. S. average = 100)	Value of land, buildings, livestock, and equipment per worker, 1945 (U.S. average = 100) [2]	Land and buildings per worker, 1945 [3]	Livestock per worker, 1945 [3]	Equipment per worker, 1945 [3]	Total cropland per worker, 1944 [4]
	Percent	Percent	Dollars	Dollars	Dollars	Acres
West North Central	168	170	7,668	1,592	926	88.1
Pacific	152	171	8,748	826	623	33.8
Mountain	143	151	6,470	1,820	719	66.5
East North Central	131	154	7,175	1,184	863	43.3
Middle Atlantic	108	97	3,942	1,018	838	23.9
New England	96	88	3,960	745	529	15.6
West South Central	71	73	3,433	616	320	33.4
South Atlantic	61	46	2,212	343	187	14.6
East South Central	51	41	1,923	351	177	16.1
United States	100	100	4,622	844	513	37.4

[1] Based on number of farm workers from Bureau of Agricultural Economics Farm Labor Reports, 1944, the production year reported in the 1945 census.
[2] Livestock includes horses and mules and all other farm animals, including poultry and bees. Equipment includes motor vehicles, machinery, and equipment.
[3] From the 1945 Census of Agriculture.
[4] Land from which one or more crops were harvested plus estimated acreage of crop failure and summer fallow.

In the fertile grain and livestock region of the West North Central States, for example, production per worker in 1944 was 168 percent of the United States average, and the 1945 value of land and buildings, livestock, and equipment per worker was 170 percent of the average. In the East South Central States, on the other hand, production per worker was only 51 percent of the United States average, and the value per worker of the indicated capital items was 41 percent of the average (table 6). The West North Central farmers had about 88 acres of cropland per worker in 1944, whereas the East South Central farmers had about 16 acres per worker. Equipment per worker averaged $926 in value in the West North Central States compared with $177 in the East South Central division. Most of these relationships among the regions probably have not changed greatly since the 1945 census was taken.

Although total acres of cropland per worker in the South Atlantic and East South Central States is just about the same as the average for New England, production per worker in New England is nearly 60 percent greater than production per worker in the South Atlantic States, and almost double the production per worker in the East South Central States. The value of equipment per worker in New England is also about three times the value per worker in the 2 southern regions.

These differences are due in part to characteristics of the agricultural divisions that make them adaptable to different types of agriculture. These characteristics make the Southern Divisions well suited for such cash crops as cotton, tobacco, and vegetables. Because of low yields of feed crops and pasture per acre, however, they are not so well adapted to specialized livestock and poultry production. New England farmers, on the other hand, are much less dependent on their "cropland" acres for production. Large permanent pastures and hay lands furnish roughage for dairy cattle during summer and winter. Grain and other concentrates with which to round out the dairy enterprise are brought in, principally from the Corn Belt. Large poultry enterprises, from which the great eastern cities are supplied with fresh eggs and poultry meat, are fed by bringing in feed grown elsewhere. Thus, the resources per farm worker in buildings, machinery, and power, are much less dictated by the cropland acres per worker in New England than in the Southern Divisions.

Westward from New England are the Central and Northern areas of less rugged topography, more productive land, and opportunities for operations on a larger scale, or for more diversity because of climatic conditions. Here even greater opportunities for greater production per worker become apparent through the use of more machines and facilities for livestock production.

More Product Per Hour of Work

Agriculture has gone hand in hand with industry in cutting the number of hours used to produce a unit of product. According to Fabricant,[4] practically every industry engaged in extraction, fabrica-

---
[4] Solomon Fabricant, Labor Savings in American Industry, 1899–1939, Occasional Paper 23, National Bureau of Economic Research, Nov. 1945.

tion, power production, transportation, and communication reduced its man-hour requirements per unit of product during the period, 1899–1939. The great majority also reduced the number of workers employed per unit. The average reduction in workers per unit of product for all these industries combined was about 58 percent. When labor input is measured by man-hours it is about 65 percent. Both percentages probably understate the decline as they do not take full account of the improved quality of product and they ignore the great economies in material and fuel consumption.

Considering our total economy, including agriculture, the decline from 1899 to 1939 in persons employed, including the self-employed, per unit of product was about 40 percent. In terms of national product per worker the increase during the 40-year period was 70 percent. This does not represent the total gain from increased productivity because hours of labor per week declined. In terms of output per man-hour, the increase in productivity was about 100 percent.

As the purpose of this discussion is to show the effect of mechanization on gross farm production and farm output per hour of farm work, several new estimates have been necessary. These have been made for periods for which basic data are available, beginning with 1909–13, and for each year from 1939 to 1945, inclusive.

TABLE 7.—*Gross farm production, farm output, and man-hour requirements, United States, selected periods and years, 1909 to 1945*

[For indexes, 1917–21 = 100]

Period or year	Gross farm production	Farm output	Man-hour requirements [1]	Man-hours per unit of—		Gross production per hour	Farm output per hour
				Gross production	Farm output		
	Index	Index	Million hours	Index	Index	Index	Index	Index
1909–13	93	93	22,262	97	104	104	96	96
1917–21	100	100	22,983	100	100	100	100	100
1927–31	107	116	22,193	97	91	84	110	120
1939	111	126	20,454	89	80	71	125	142
1940	113	130	20,412	89	79	68	127	146
1941	117	136	20,617	90	77	57	130	151
1942	129	150	21,132	92	71	61	140	163
1943	126	147	21,026	91	73	63	137	160
1944	131	154	21,182	92	70	60	142	167
1945	130	153	20,655	90	69	59	144	170
Percentage change:								
1917–21–1939	11	26	−11		−20	−29	25	42
1939–44	18	22	3		−12	−15	14	18
1939–45	17	21	1		−14	−17	15	20

[1] See footnote 2, table 5.

During the war and postwar years of 1917–21, an average of nearly 23 billion man-hours were required annually in the operation of farms of the United States. In the later war years, 1942–45, an average of about 21 billion hours were required. While this decrease of about 9 percent in total man-hours was occurring, gross agricultural production increased 29 percent, and farm output for human use increased 51 percent (table 7). The combination of increased production with fewer man-hours meant that during World War II it took only 71 percent as many man-hours to produce a unit of gross production as during the period 1917–21. Looked at another way, 1 hour of man labor produced about 41 percent more gross product in 1942–45 than in 1917–21. The increase in farm output for human consumption per hour of man labor was considerably greater, amounting to 65 percent.

Different degrees of change are apparent when these changes are considered separately for the period, 1917–21 to 1939, and for the period 1939–45 (table 7). The 21-year inter-war period was an extended period of farm mechanization, and a large part of the increase in gross production per hour was caused by increases in power machines for field and road work, as is shown later. From 1939 to 1945, gross production increased considerably more than during the previous 21 years. But total man-hours required in farm operation did not decrease—instead they increased 1 percent. Gross production per man-hour increased 60 percent as much in this short period as the increase in the previous 21 years. Increased yields per acre and per animal, as well as increased mechanization, were influential in increasing production per hour of labor, especially during World War II.

TABLE 8.—*Man-hour requirements per unit of production for selected farm products, United States, selected periods and years, 1917–44*
[Index numbers (1917–21 = 100)] [1]

Product	1917–21	1927–31	1939	1942	1943	1944	Percentage change	
							1917–21 to 1939	1939 to 1944
Milk	100	87	86	81	80	78	−14	−9
Meat animals	100	93	85	83	81	78	....	−8
Poultry	100	94	90	85	82	85	−10	−6
Total meat animals and animal products	100	91	88	81	78	78	−12	[2] −11
Feed and food grains	100	85	74	58	64	60	−26	−19
Cotton	100	84	66	62	62	58	−34	−12
All truck crops	100	85	79	77	79	73	−21	−8
Vegetables except truck	100	78	74	65	66	67	−26	−9
Total crops	100	90	76	66	69	66	−24	−13
Gross farm production	100	91	80	71	73	70	−20	−12
Farm output	100	84	71	61	63	60	−29	−15

[1] 1919–21 production used; data not available for 1917–21.
[2] The percentage decrease in hours per unit of production of all livestock was larger than for any class of livestock shown because of changes in the relative importance of the individual livestock items. Poultry and meat-animal production, which requires less than one-half as much labor per unit of product as does milk, increased much more than milk production from 1939 to 1944.

The largest decrease in man-hours per unit of production was for cotton from 1917–21 to 1944, and for feed and food grains from 1939 to 1944 (table 8). In the first instance the decrease was due largely to increased cotton yields, which required little more labor per acre for preharvest work, and to some increase in mechanization in preharvest and harvest work. Cotton yields averaged 163 pounds per acre in 1917–21 when yields were low because of excessive boll weevil damage, and

238 pounds per acre in 1939 when the greatly reduced cotton acreage consisted of the more productive lands. Food-grain production was mechanized considerably before World War II. The practice of combine-harvesting, or "combining," instead of binding and threshing food grains continued to expand in World War II, and hours per unit of corn were cut drastically by greater use of mechanical corn pickers and by picking corn from the standing stalk in areas that previously harvested by cutting, shocking, and hand husking from the shock. Hours per unit of production of truck and vegetable crops were reduced about as much as they were for feed and food grains in the period 1917–21 to 1939. Preharvest jobs for these crops, and to some extent harvest jobs, were mechanized during that period.

TABLE 9.—*Estimated "labor savings" in farm production resulting from changes in mechanization, crop yields, and other factors since 1917–21, United States*

Item	Man-hours saved in *1944* production because of changes in methods and practices since 1917–21		Man-hours saved in *1939* production because of changes in methods and practices since 1917–21		Man-hours saved between 1939 and 1944 because of changes in methods and practices since 1917–21	
	Millions	*Percent*	*Millions*	*Percent*	*Millions*	*Percent*
Savings because of *increased mechanization* including decrease in horse and mule numbers:						
Totals	4,200	48	3,000	60	1,200	30
Averages per year	168		150		240	
Savings of preharvest and harvest labor because of *increased yields*:						
Totals	1,700	19	850	17	850	21
Averages per year	68		42		170	
Savings in livestock labor because of *increase in size of enterprise, increased production per animal, etc.*:						
Totals	1,200	13	550	11	650	16
Averages per year	48		28		130	
Savings because of *spreading* of overhead labor over larger volume of production:						
Totals	650	7	100	2	550	14
Averages per year	26		5		110	
Savings because of shifts in importance of *labor intensive and labor extensive enterprises:* [1]						
Totals	(600) [2]	(7) [2]	(600) [2]	(12) [2]	0	0
Averages per year	(24) [2]		(30) [2]		0	
Miscellaneous savings because of *change in method, elimination of operations, work simplification,* increase in some custom-hired operations, etc.:						
Totals	1,850	20	1,100	22	750	19
Averages per year	74		55		150	
Totals	9,000	100	5,000	100	4,000	100
Averages per year	360		250		800	

[1] Production of enterprises in 1919–21 used in making estimates. Data not available for 1917 and 1918.

[2] Increases instead of savings. To be subtracted from the sum of the other items.

Reduction in the number of man-hours required to produce a unit of meat animals and animal products was only half as great from 1917-21 to 1939 as the reduction in hours per unit of crop production—12 percent and 24 percent respectively (table 8). In the 6 wartime years, 1939-44, the decrease in hours per unit of livestock production was nearly as large as the total decrease for the 20 years from 1917-21 to 1939. More milking machines and better barn and barnyard facilities and equipment helped to reduce man-hours per unit of livestock production. But higher milk production per cow, higher egg production per hen, and higher hog, cattle and sheep production per unit of breeding stock probably were also important, especially during the war.

Decreases in the hours per unit of farm output were somewhat larger than the decreases for gross farm production, because of the continued decrease in horse and mule numbers on farms since 1918 (table 8).

If all farmers in the United States had farmed in 1944 as they did in 1917-21—that is, had used the same machines and tools and the same power units—and produced their crops and livestock in about the same proportions and at the same rates of yield, the larger volume of gross production in 1944 would have required 9 billion more man-hours than were actually required (table 9).[5] Without this saving in man-hour requirements our 1944 farm production would have required about 30 billion man-hours instead of 21 billion hours. In terms of man-years of 2,000 hours each, 4½ million more farm workers would have been needed to do the 1944 production job if our farmers had

---

[5] In 1944, gross farm production amounted to 11.2 billion dollars at average 1935-39 prices. Counting 100 dollars as a unit of production, gross production in the last full year of war totaled 112 million units. About 189 man-hours of farm labor (including farm-maintenance labor) were required for each unit of production in 1944. In 1939, about 215 man-hours were required for each of the 95 million units of production. Gross production averaged 86 million units in the period 1917-21 and each unit required 269 man-hours.

Total labor savings, as used in this publication, are a function of reduced man-hours per unit of production and total volume of production in the year for which total savings are estimated. Thus in 1944 each unit of production required about 80 fewer hours of labor than in the 1917-21 period; a saving of 80 hours per unit applied to the 112 million units produced in 1944 results in total savings of approximately 9 billion hours. This means that if the 1944 volume of production had been produced with the methods and practices, and at the same level of yields, as characterized the 1917-21 period, 9 billion additional man-hours of labor would have been required. Stated in another way, if total man-hour requirements had increased from 1917-21 to 1944 by the same percentage as total units of production increased, total man-hour requirements in 1944 would have been 9 billion more than actually was the case.

With the practices, methods, and yields that prevailed in 1939, each unit of production required 54 fewer hours than in 1917-21. This saving of 54 hours per unit applied to the 95 million units of production in 1939 resulted in a total labor saving of approximately 5 billion hours. Again this means that if total man-hour requirements had increased by the same percentage as gross production increased from 1917-21 to 1939, total man-hours in 1939 would have been about 5 billion more than actually were required in 1939.

The difference between the 9 billion hours of savings for the entire period, 1917-21 to 1944, and the 5 billion hours of savings from 1917-21 to 1939 is 4 billion hours. This means that 55 percent of the total saving of 9 billion hours had already been attained by 1939, and that 45 percent of the total savings came between 1939 and 1944.

In summary, savings for the entire 25-year period and the two sub-periods are measured from a common base of 1917-21. From this common base period to 1944 there was a saving of 80 man-hours per unit of production on 112 million units produced in 1944. Most of this saving, or approximately 5 billion hours, had occurred by 1939 when 95 million units were produced with 54 fewer hours per unit than would have been required under 1917-21 production conditions.

been using the farm power, machines, and production practices that were in common use in 1917–21.

A combination of several other developments has been even a little more effective than farm mechanization in reducing man-hours per unit of farm production. Savings in man-hours, as defined above, because of farm mechanization, from 1917–21 to 1944 amounted to 48 percent of total savings from all developments. The remaining 52 percent of the total was composed principally of savings because of increased production per acre of cropland, increase in size of livestock enterprise, increase in production per animal, spreading of overhead over a larger business, and several other factors. These others included changes in methods of handling enterprises, elimination of operations, work-simplification features, and increase in some custom-hired operations. Total savings from these items plus those from mechanization amounted to 9.6 billion hours. But the net savings were only 9 billion hours because of an increase of 600 million hours as a result of shifts in production to more intensive labor-using crops and livestock.

Private and governmental research, and farmer experimentation and adoption, have been responsible for technological developments that have brought about more production from each hour of farm work. Engineers, plant breeders, animal breeders, agronomists, pathologists, chemists, entomologists, and others have contributed to increased farm mechanization and increased production per acre and per animal. Both mechanization and increased production have increased production per hour of labor.

TABLE 10.—*Approximate importance of various types of machines responsible for "labor savings" in agriculture, 1917–21 to 1944*

Item	Approximate increase in number of machines on farms 1917–21 to 1944	Approximate number of man-hours saved by increase in mechanization, 1917–21 to 1944
	Thousands	*Thousands*
Tractor-operated plows, listers, middlebusters, disk harrows, field tillers, cultivators, row-crop planters, and grain drills	5,200	380,000
Tractor-operated grain combines, corn pickers, grain binders, and row-crop binders	900	400,000
Tractor-operated mowing machines and pick-up balers	450	60,000
Tractor-operated miscellaneous equipment		100,000
Farm automobiles and motortrucks	3,400	1,400,000
Manure spreaders, small tools and gadgets, increase in size of horse-drawn equipment in some areas, increase in machine over hand work in planting, hay-making, etc., better and more convenient gates, fencing, storage, etc		700,000
Milking machines	300	220,000
Other savings on meat animals and animal products		60,000
Net savings in chore work on horses and mules because of displacement by mechanical equipment (horse and mule chores, minus farm maintenance work on tractors, autos, trucks, and other farm machinery and equipment)		880,000
Total		4,200,000

PROGRESS OF FARM MECHANIZATION

Of the 4.2 billion man-hours saved because of farm mechanization, 1.4 billion, or one-third, were saved by an increase of 3.4 million farm automobiles and trucks (table 10). These transportation vehicles, with the good roads that accompanied their increase in numbers, enable farmers to haul livestock and other farm products to market and to haul supplies home, in only a fraction of the time it took when livestock were moved on foot and horse-drawn vehicles and dirt roads were the rule. Savings in hauling on the farm have been made. On the average, each automobile or truck saves the farmer more than 400 man-hours a year, compared with the time it would take to do the same hauling with horses and mules. As only one-half of the automobile use was considered farm business and so was included in these computations, other worth-while savings have been made in the time spent on personal matters.

Horses and mules disappeared between 1917-21 and 1944 by more than 13 million head. Savings in chores involved in caring for these animals amounts to 880 million man-hours, after deducting the additional farm time for caring for the tractors, motor vehicles, and other machinery that displaced them. A part of this saving has come from the fact that more servicing of tractors, automobiles, and trucks is now done by town and city workers whereas chore work on farm horses and mules has always been done on the farms.

The actual decrease in chore labor on horses and mules during the 25-year period approximated 1,130 million man-hours. This decrease was brought about by the disappearance of 13 million head of horses and mules from farms and by a decrease in the time spent in caring for each animal. Not all of this gross saving in man-hours was available for producing food and fiber for human use because of an increase of 250 million farm man-hours used in caring for more motor vehicles and other farm machinery. The difference between the two, or the net saving, was 880 million hours. Farm maintenance of automobiles and motortrucks required 145 million of the 250 million hours, and farm maintenance of tractors and other machinery required the remainder, or 105 million hours. About 760 of the total net savings of 880 million hours was attributed to the added tractors and tractor equipment, and the remainder, or 120 million hours, to the fast-growing numbers of farm trucks and automobiles.

In 1944, tractor-operated equipment for field operations—such as seedbed preparation, seeding, cultivating, and harvesting crops—and for doing miscellaneous farm jobs reduced man-hours below those which would be required with teams, by 940 million hours. This, added to the net decrease of 760 million hours used in caring for fewer horses and mules, amounts to 1.7 billion hours. As tractors on farms increased from 1917-21 to 1944 by about 2 million, the yearly saving per tractor has been about 850 man-hours.

Milking machines added to farm equipment over the 25-year period saved farmers 220 million hours in 1944. Manure spreaders, and the numerous small tools and conveniences introduced on farms since 1917-21, lowered man-hours in 1944 from what they would have been if farmers were still farming as they were in 1917-21 by 700 million.

Most of these savings in time because of mechanization, (55 percent) have been made in growing, harvesting, and hauling crops. Twenty-six percent of the total was reduced chore work on horses and mules, 15 percent on meat animals and animal products, and 4 percent of the savings because of mechanization were on general maintenance of farm plant.

Savings in labor because of increased yields from 1917–21 to 1944 amounted to 1.7 billion hours (table 9). Most of this saving was in the form of preharvest crop labor. If acre yields had remained as they were in 1917–21, farmers would have needed about 69 million more average acres to produce the 1944 volume of production than were actually used. Seedbed preparation, seeding, and cultivating these 69 million acres with 1944 equipment and power would have taken about 1.5 billion more man-hours than were required on the smaller acreage actually used. Nearly 85 percent of these 1.5 billion preharvest hours on the large acreage would have been for cotton, corn, truck crops, and tobacco; nearly 50 percent would have been for cotton alone. Thus, these crops were largely responsible for the savings in hours of preharvest work due to increase in yields. The percentage increase in yields per acre, 1944 over 1917–21 average, were about as follows: Cotton, 80 percent; tobacco, 35 percent; wheat, 35 percent; and corn 20 percent. All crop production per acre was about 25 percent higher in 1944 than in 1917–21.

About 170 million hours were saved in the harvest work of 1944 production because of the increase in yields of crops over the 1917–21 average. Particularly in the case of crops whose harvest operations are mechanized, an increase in yield does not result in a proportional increase in hours required for harvesting—it results in decreased harvest-labor requirements per unit of production. Relatively less saving in harvest labor per unit results from increased yields of such crops as cotton, tobacco, or truck crops, whose harvest involves primarily hand labor. Of the total savings in harvest labor due to increased yields from 1917–21 to 1944, about three-fourths were made on the wheat and corn crops.

The volume of meat animals and animal products that was produced in 1944 would have required 1.2 billion more man-hours than were required had it not been for the increase in size of livestock enterprises and the increase in production per animal compared with the 1917–21 average. There were more milk cows per farm in 1944, and milk production per cow had increased 21 percent during the period. A decreasing number of man-hours are required to produce 100 pounds of milk as cow numbers in the herd increase and as production per cow increases.

Egg production has become more specialized, and egg production per hen has increased during the same period by 27 percent, and considerably less labor is now required to produce a dozen eggs. Commercialized broiler and turkey production in recent years has helped to reduce the man-hours used to produce a hundred pounds of chicken and turkey meat. More hogs per farm, better sanitation, and more attention to saving pigs have reduced losses of both pigs and more mature animals. Housing, feeding, and watering conveniences mean fewer hours required to produce 100 pounds of hogs. It takes less labor to produce a given quantity of beef, lamb, and wool than it did 25 years ago. All combined, production of meat animals and animal products has increased much faster than have the numbers of breeding stock. (See table 37 in the appendix).

The volume of agricultural production increased at a more rapid rate between 1917–21 and 1944 than did general farm-maintenance labor or overhead, and production in 1944 required 650 million fewer overhead hours than the same volume of production would have required in 1917–21 (table 9). More production per hour of labor spent in maintenance and repair of buildings, fences, and machinery accounted for most of the overhead savings in labor per unit of farm production.

About 100 million units of agricultural production, exclusive of product added by horses and mules, pastures, and a few miscellaneous crops, were produced in 1944.[6] If this 1944 production had been of the same composition as the production of 1917–21, it would have taken 160 direct hours, exclusive of indirect farm-maintenance labor, to produce one unit of 1944 production. Actually, because of change in composition, 166 direct hours were required in 1944 to produce a unit of production. Thus, because of change in composition, it took 6 more hours per unit of production, or a total of 600 million more hours to produce the 100 million units of production.

Two groups of products contributed to the plus side of the net increase of 600 million hours. One group—composed of milk, truck crops, and tobacco—whose production increased more than the average and whose labor requirements per unit of production were larger than the average for all commodities, contributed a plus of 387 million hours (table 11). Milk was the most important contributor in the group. This was due to the relatively large increase in milk production—60 percent—compared with the average increase of about 36 percent for all commodities, and to the high labor requirements per unit of product—398—compared with 160 for all products.

The second group consists of food and feed grains, hay, and vegetables, excluding truck. This group contributed 471 million hours to the plus side of the total. But unlike the first group, production of the commodities in group 2 increased less than average, and the commodities required less than average amounts of labor per unit of production. The influence here is indirect in that the lower rate of increase made available more of all production resources for commodities that required above-average labor per unit of production. Food and feed grains combined made the dominating item in this group. Food and feed grains production increased from 1917–21 to 1944 by only 14 percent, compared with the average increase for all commodities of 36 percent, and only 111 hours of labor were required in 1944 to produce 1 unit of production, compared with an average of 160 hours for all commodities.

The next two groups of commodities shown in table 11 contributed to the minus side of the net increase of 600 million hours. Cotton and sugar crops require much labor per unit of production but neither crop increased in production as much as the average increase of all commodities. Cotton production from 1917–21 to 1944 increased only

---

[6] One hundred dollars' production in terms of 1935–39 average prices equals one unit of production. All of the labor savings through change in numbers of farm horses and mules were attributed to mechanization. "Product added" by horses and mules amounted to 3 million units of production in 1944. About 9 million units of production in the form of pastures and a few miscellaneous crops were omitted from this analysis because necessary data are lacking.

about one-third as much as the average, and production of the sugar crops decreased by 13 percent.

The second group that lowered the average labor requirements per unit of production consisted of meat animals, oil crops, poultry products, and fruits and nuts. Each of these require less than the average labor per unit of product, and their production increased at a much higher rate than the average increase for all commodities.

TABLE 11.—*Effect on "labor savings" of shifts in composition of production, 1917–21 to 1944*

Item	Labor requirements per unit of production, 1944 [1]	Increase in production, 1917–21 to 1944 [2]	Composition of production		Contribution to total "labor savings" because of change in relative composition
			1917–21 average [2]	1944	
	Hours	Percent	Percent	Percent	Million hours
Enterprises that raised average labor requirements per unit of all production:					
Enterprises with above-average requirements:					
Milk [3]	398	60	7.7	9.0	350
Truck crops	169	128	4.0	6.7	25
Tobacco	209	43	3.3	3.5	12
Enterprises with below-average requirements:					
Feed and food grains	111	14	40.7	34.1	338
Hay	105	14	10.1	8.5	100
Vegetables, except truck	88	23	4.2	3.8	33
Enterprises that lowered average labor requirements per unit of all production:					
Enterprises with above-average requirements:					
Cotton	274	13	8.7	7.2	−166
Sugar crops	252	[4] −13	1.1	.7	−38
Enterprises with below-average requirements:					
Meat animals [3]	144	53	9.8	10.9	−19
Oil crops	151	647	.5	2.6	−19
Poultry products [3]	156	90	5.9	8.2	−9
Fruits and nuts	151	63	4.0	4.8	−7
Average or total	[5] 160	36	100.0	100.0	+600

[1] The unit of production is 100 average 1935–39 dollars.
[2] Averages for 1919–21 were used; data not available for 1917 and 1918.
[3] Production of livestock and livestock products was expressed in terms of "product added."
[4] A decrease.
[5] Weighted average, assuming that same composition of production existed in 1944 as in 1919–21.

Between 1917–21 and 1944 farmers reduced the hours required per unit of production in other ways. For example, a large part of the topping and stripping of blades from the corn stalks in the South for "bundle feed" has disappeared with the expansion of hay and other forage-crops. In some areas, there is less hand hoeing of corn and other row crops. Farm and building conveniences that save work are more prevalent. Commercial haulers of grain, livestock, fruit, and other farm commodities have helped to reduce man-hours per unit of production. Prices paid for important supplies—as fertilizer, lime, feed, and building supplies—now frequently include delivery at the farm. Miscellaneous farm labor savings of this sort over the 25-year period are estimated at about 1.8 billion hours for a volume of farm production equal to that of 1944 (table 9).

Relatively large parts of the changes that have been described took place during the war period, 1939–44. For example, if all farmers had been farming in 1939 in the way they were in 1917–21, the 1939 volume of production would have required 5 billion more man-hours (table 9). This means that 4 billion of the 9 billion hours of "labor savings" were saved after 1939, and were possible principally because of increased yields, increased size of livestock enterprises, spreading of overhead over a larger volume of business, and increased use of farm machines. On a percentage basis, mechanization contributed less to labor savings during the war period, 1939–44, than in the longer interwar period, 1917–21 to 1939. In terms of hours per year, however, the savings were much larger during the war than in the earlier period.

TABLE 12.—*Estimated "labor savings" between 1939 and 1944 resulting from changes in farming methods and practices since 1917–21 and since 1939, United States totals*

Item	Man-hours saved between 1939 and 1944 by changes in methods and practices since 1917–21		Man-hours saved between 1939 and 1944 by changes in methods and practices since 1939	
	Million hours	Percent	Million hours	Percent
Savings because of *increased mechanization*, including decrease in horse and mule numbers	1,200	30	850	28
Savings of preharvest and harvest labor because of *increased yields*	850	21	800	27
Savings in livestock labor because of *increase in size of enterprise, increased production per animal*, etc.	650	16	550	18
Savings because of *spreading of overhead labor over larger volume of production*	550	14	400	13
Savings because of shifts in importance of *labor intensive and labor extensive enterprises*	0	0	200	7
Miscellaneous savings because of *change in methods, elimination of operations, work simplification*, increase in custom-hired operations, etc.	750	19	200	7
Total	4,000	100	3,000	100

If the labor savings from 1939 to 1944 were measured from the base year 1939 instead of the base period 1917–21, the total would be only three-fourths as large, or 3 billion man-hours (table 12). Labor savings from increased mechanization would account for 28 percent of the total instead of 30 percent, and labor savings because of increased yields would be 27 percent instead of 21 percent. Wartime shifts to less labor-intensive crop and livestock enterprises brought greater savings during the war than in the prewar period.

More Food and Fiber for Human Consumption

The shift from animal power toward mechanical power on farms and in cities, towns, mines, etc., that started after World War I and continued through World War II has resulted in two important accomplishments. It has helped to reduce the farm-labor force needed and the

TABLE 13.—*Acreages of harvested crops used for specified purposes, United States, 1910–45*

Crop year beginning in	Crops harvested [1]	Acreages used for producing export products [2]	Acreages used for producing—				Total population July 1 [5]
			Feed for horses and mules [3]		Food, fiber, and tobacco for domestic consumption [4]		
			On farms	In cities, mines, etc.	Total	Per capita	
	Million acres	*Million acres*	*Million acres*	*Million acres*	*Million acres*	Acres	*Million*
1910	325	36	70	16	203	2.21	92
1911	330	40	72	15	203	2.16	94
1912	329	41	73	15	200	2.11	95
1913	333	42	74	15	202	2.08	97
1914	334	55	76	14	189	1.91	99
1915	340	47	77	14	202	2.00	101
1916	340	51	77	13	199	1.95	102
1917	349	42	78	12	217	2.11	103
1918	362	60	79	11	212	2.02	105
1919	364	54	79	10	221	2.10	105
1920	360	58	77	10	215	2.03	106
1921	359	64	77	8	210	1.93	109
1922	355	48	76	7	224	2.04	110
1923	354	46	76	6	226	2.02	112
1924	355	52	74	5	224	1.96	114
1925	360	43	72	4	241	2.08	116
1926	359	52	70	4	233	1.99	117
1927	358	48	68	3	239	2.01	119
1928	361	48	66	2	245	2.02	121
1929	365	42	64	2	257	2.11	122
1930	369	38	61	2	268	2.18	123
1931	365	35	60	1	269	2.17	124
1932	371	34	58	1	278	2.22	125
1933	340	27	56	1	256	2.03	126
1934	304	19	55	1	229	1.83	125
1935	345	20	53	1	271	2.13	127
1936	323	18	51	1	253	1.98	128
1937	347	28	50	1	268	2.08	129
1938	349	21	46	1	281	2.16	130
1939	330	23	43	1	263	2.01	131
1940	339	15	41	1	282	2.14	132
1941	342	15	38	1	288	2.17	133
1942	346	22	37	1	286	2.12	135
1943	355	37	37	1	280	2.06	136
1944	359	33	36	1	289	2.09	138
1945	355	35	34	1	285	2.04	140

[1] Area in 52 principal crops harvested or estimated equivalent plus acreages in fruits, tree nuts, and farm and market gardens.

[2] Crop exports from 1910 to 1939 are based on yields of specified year applied to gross exports for year beginning July 1, or month representing beginning of crop season. Acreages for livestock exports from 1910 to 1939 are based on average crop yields for 1935–39, and are for the year beginning July 1. Acreages for exports and lend lease from 1940 to 1945 for both crops and livestock are based on 1940–43 average crop yields.

[3] Feed computations for horses and mules are based on United States average yields of corn, oats, and all hays. From 1910 to 1919 the calculations allow 800 pounds of oats, 1,600 pounds of shelled corn, and 1.8 tons of hay per head for farm horses and mules 3 years old and over, and animal-unit equivalents for younger animals. Beginning with 1920, it was assumed that the rate of feeding corn declined 10 pounds per head annually and the rate of feeding hay increased 20 pounds. For nonfarm horses and mules the quantity of grain and hay fed per head annually was estimated to average about one-third more than for farm horses and mules.
[4] Includes products used by our military forces in this country and abroad, and by our domestic civilian population.
[5] Includes persons in our military forces in this country and abroad.

time required to produce a unit of farm product, and it has diverted land and other resources from the production of feed for horses and mules to the production of food for human consumption. Nine million crop acres were thus released during the 6 high-yielding wartime years alone, and 55 million crop acres were so released during the 27 years since horse and mule numbers reached a peak in 1918 (table 13). The total released acreage would have been larger if yields of feed crops had not increased so substantially during World War II. Feed from large acreages of pasture land has been diverted also from maintenance of workstock to the production of livestock food for human use.

After allowing for acreages used to grow products for export, lend lease, and feed for horses and mules, the acreages for producing products for domestic consumption have averaged, during 1940–45, about 2.10 acres per capita, varying annually to some extent because of variations in crop yields. During the 1940–45 harvest seasons production per acre of cropland averaged considerably above the 1935–39 average. In addition, the per capita use of these higher yielding acres was somewhat greater than the average acreage used in 1935–39.

TABLE 14.—*Changes in acreages of harvested crops used for specified purposes, 1918–22 to 1940–44, United States averages*

Crop acreages and population	1918–22 averages	1940–44 averages	Change from 1918–22 to 1940–44	
			Number	Percent
Crop acreages harvested:				
Total..........million acres..	360.0	348.2	−11.8	−3
For export crops..........do...	56.8	24.4	−32.4	−57
For horse and mule feed...do...	86.8	38.8	−48.0	−55
For domestic use..........do...	216.4	285.0	+68.6	+32
For domestic use				
..........acres per capita..	2.0	2.1	+.1	+5
Population..........millions..	107.0	134.8	+27.8	+26

The wartime per capita acreages listed in table 13 include the acreages used to grow the tremendous wartime supplies of food, fibre, and tobacco that were used by our armed forces and in the making of munitions Some of the acreage was used for supplies that were diverted from military stocks to allied military and civilian uses, and to feeding civilians in occupied countries and prisoners of war. Large losses of agricultural products that occured in transit and the accumulation of military stocks before the ending of the war are also included in the per capita acreages. If correct allowances could be made for all these "exceptional" disappearances of food, fiber, and tobacco during the war, the average crop acreage per capita would be reduced to something less than 2. But owing to the greater wartime crop and livestock yields, per capita consumption actually averaged 7 to 8 percent higher than that of 1935–39.

Horse and mule numbers in the United States increased until 1918, and acreages used for growing grain and forage for them were at their height at that time. Thereafter, their continuous decline has released more and more crop and pasture land for producing food, fiber, and tobacco for market. Comparisons of the two 5-year periods, 1918–22 and 1940–44—both periods of high farm and industrial activity—show the sources of increased production to feed and clothe our increasing population (table 14). During 1940–44 total population in the United States averaged 26 percent larger than during 1918–22, and on an average each person in the later period consumed at least 10 percent more farm products than in the earlier period. During the same time harvested crop acreages for *all uses* decreased about 3 percent.

There are three noteworthy reasons why our increased population has been fed and clothed better from less land. In the first place, crop yields were considerably higher in the recent war years than in the period, 1918–22. In the second place, fewer acres were needed to grow horse and mule feed. In the third place, fewer acres were needed to produce our export and lend lease products in 1940–44 compared with 1918–22. In terms of percentages, more than 50 percent of the increased production used by our larger population is explained by larger crop and livestock yields, about 30 percent by a decrease in crop acreages required for feeding farm and off farm horses and mules, and about 20 percent in decreased acreages required for producing products for export or lend lease.

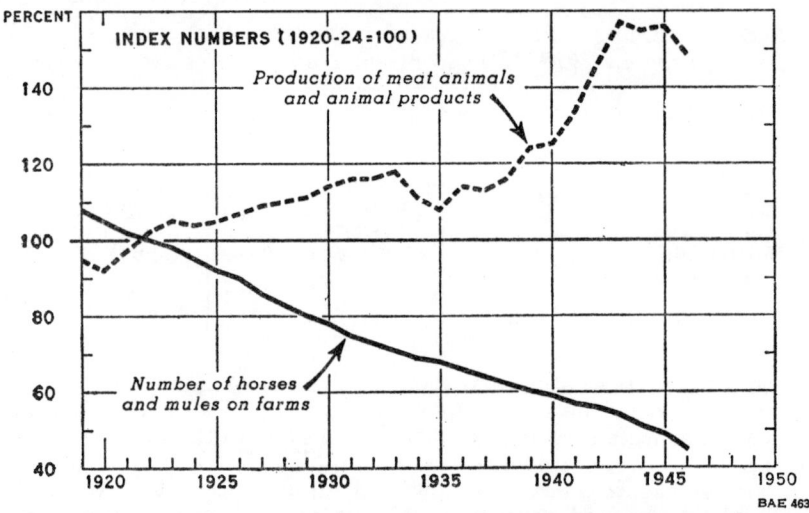

FIGURE 3—HORSE AND MULE NUMBERS ON FARMS, AND VOLUME OF PRODUCTION OF MEAT ANIMALS AND ANIMAL PRODUCTS, UNITED STATES, 1919–46.

Feed supplies released because of the decline in horse and mule numbers have contributed to the increase since 1919 in supplies of meat animals and animal products for human consumption.

Although crop yields and the composition of crop and livestock production have varied over the 35-year period, the general tendency since 1918 has been for the production of meat animals and animal products, including poultry and dairy, to increase somewhat in proportion to the decrease in horse and mule numbers (fig. 3).

The dip from normal in the production trend of meat animals and animal products which occurred in the 1930's was caused by severe drought in large areas and by Government restrictions of livestock production because of depression prices. Ever-normal granaries of feed crops that were built up in the periods of low demand for meat, eggs, butter, and milk were used later on in the succeeding war period to aid in providing exceptionally large supplies of livestock and livestock products for use both here and abroad.

The cost of raising and keeping farm work animals represents farm-produced power. Measured in 1935–39 average dollars, the volume of farm-produced power reached a peak of nearly 2 billion dollars in 1918 and then fell steadily to less than one-half of that volume in 1945. By 1945 farm-produced power made up only 7.4 percent of total gross farm production, compared with about 21 percent of the total gross production realized during 1910–14 (table 15).

TABLE 15.—*Gross farm production and farm-produced power, specified periods and years, United States, 1910–45*

Period	Gross farm production [1]	Farm-produced power [2]	
		Amount	Proportion of gross production
	Million dollars [3]	*Million dollars* [3]	*Percent*
Average:			
1910–14	8,105	1,714	21.1
1915–19	8,522	1,861	21.8
1920–24	8,758	1,713	19.6
1925–29	9,103	1,453	16.0
1930–34	8,735	1,222	14.0
1935–39	9,060	1,083	12.0
1940–44	10,567	947	9.0
1940	9,724	1,002	10.3
1941	10,052	977	9.7
1942	11,045	947	8.6
1943	10,796	910	8.4
1944	11,206	868	7.7
1945	11,145	823	7.4

[1] Gross farm production measures calendar-year production of all crops, pasture consumed by all livestock, and the product added in the conversion of feed and pasture into livestock and livestock products for human use and into farm-produced horse and mule power.
[2] Farm-produced power is the cost, in average 1935–39 dollars, of raising and maintaining farm horses and mules.
[3] Average 1935–39 dollars.

Thus, the land, labor, and equipment formerly used to produce more than 1 billion dollars worth of animal farm power (1935–39 dollars) became available during World War II for producing commodities for human consumption. Farm output, or roughly, that part of gross farm production that is available for human use, however, increased much more during the war than farm-produced power decreased. Greater use of fertilizers and improved strains and varieties of seeds, and favorable weather, brought unusually good crop yields and more abundant feed and pasture crops for livestock. But owing to the large supplies of labor-saving machines on farms, and even with considerably less farm labor, these larger wartime crops were harvested, stored, and hauled to market with little loss from spoilage.

Less Hand Labor Needed

Farming requires hard work, much of which is done with the hands or with small hand tools. Of the 21.2 billion man-hours spent on farms in 1944, approximately 60 percent, or 13.0 billion hours were done with the hands or with small hand tools—ax, pitchfork, shovel, hoe, husking peg, pruning knife. It is significant that this high percentage of hand labor remains after a hundred years of substitution of mechanical or animal power for human power (table 16).

Further mechanization of farm jobs will reduce the hand work used in agriculture. It may eliminate another 2 or 3 billion hand hours within the next decade, providing mechanical cotton pickers come into general use, and providing the general trend of eliminating hand jobs during the last 10 years is speeded up. This total seems now to be an outside possibility, with considerable chance that the actual elimination will be appreciably less. An elimination of 2 or 3 billion hours of hand labor probably would mean a net reduction of one-third to one-half as many hours. A part of the direct savings in labor occasioned by changing from hand tools to power tools is offset by additional hours used in caring for the more complicated tools and machines. In exceptional cases the total hours per unit of product may be almost as great with machines as with hand tools; the real advantages then come from doing the job more quickly and with less effort.

The greatest need for hand work is in the care of livestock. At present, 75 percent of the man-hours in livestock work on farms is hand labor, and more than one-third of all work on farms is caring for livestock (table 16). A further reduction of 17 percent in the hours of hand work spent on livestock would result in a decrease of 5 to 8 percent in all man-hours now spent on livestock on farms. To eliminate more hand labor in livestock care would mean more and better feeding and watering conveniences, more mechanization in handling manure in the barns, more use of milking machines, etc. With greater coverage in rural electrification, these things will be possible. There will remain, however, the 2 to 3 million small farms with only a few head each of the different kinds of livestock. Here, electrically operated water pumps will eliminate a lot of hand labor in watering the livestock but hand milking will continue on most of the 3.5 to 4 million farms that have 9 or less cows per farm.

In 1944, nearly one-half of the 21 billion hours spent in farm work were devoted directly to crops. Nearly one-half (47 percent) of these hours was hand work. It was relatively heavy on such crops as fruits, vegetables, tobacco, cotton, potatoes, and peanuts. Thirty-one percent of the 2.6 billion man-hours used in growing and harvesting corn were hand-hours, and even in wheat production nearly one-fourth of the man-hours in 1944 was hand labor.

TABLE 16.—*Estimated total number of man-hours, and hours of hand labor used on farms in 1944, and possible reduction in hours of hand labor in next 10 years*

Item	Total number of man hours, 1944	Hours of hand labor [1]			
		Hours in 1944		Possible hours by 1954 [2]	Possible reduction in hand labor, 1944-54
		Number	Percent of total hours		
	Millions	*Millions*	*Percent*	*Millions*	*Percent*
Wheat	500	115	23	75	35
Other food grains	88	22	25	16	26
Corn	2,552	791	31	511	35
Other feed grains	734	183	25	122	33
Hay and grass seeds, exclusive of grain hays	918	314	34	230	27
Peanuts	240	110	46	72	34
Soybeans and flaxseed	157	24	15	16	33
Tobacco	733	527	72	366	31
Cotton	2,001	1,101	55	560	49
Potatoes	198	103	52	85	17
Dry beans and peas	50	6	12	5	17
Other vegetables	1,244	837	67	628	25
Sugar beets and cane	115	71	62	52	27
Sorgo, cane, and maple sirup crops	71	53	75	46	13
Fruits and nuts	741	563	76	460	18
Miscellaneous crops	94	55	59	37	33
Total crops	10,436	4,875	47	3,281	33
Dairy cows	3,640	2,657	73	2,111	21
Other cattle	741	430	58	370	14
Hogs	623	486	78	362	25
Poultry and eggs	1,430	1,124	79	981	13
Sheep, lambs, and wool	237	190	80	166	13
Horses and mules	862	776	90	690	11
Total livestock	7,533	5,663	75	4,680	17
Farm maintenance	3,212	2,474	77	2,249	9
Grand total	21,181	13,012	61	10,210	22

[1] Hours of hand labor, as used here, are hours worked with hands or with hand tools. The cutting and trimming, loading and unloading of trees for firewood is hand labor. Cutting the wood into stove lengths with a crosscut saw is hand labor; with a buzz saw, machine labor. Driving a wagon in hay harvest is not hand labor, but pitching on the hay or loading after a hay loader, is hand labor. Loading and unloading manure with a pitchfork is hand labor; if a manure spreader is used the unloading is not hand. Operating a pick-up hay baler is not hand labor, but the men who pick up the bales in loading them are doing hand labor. Much hand work is used in shocking, hauling, and threshing grain; little is used in combining grain.

[2] This date is used merely as a convenience in indicating a lapse of 10 years.

Further mechanization of harvesting small grains in the East, cotton, hay, sugar beets, potatoes, and corn, will reduce hand work substantially, and as the machines used in harvesting most of these crops are labor savers, mechanization will also reduce the total man-hours. It may be possible in the next 10 years to cut the hand labor used in crop production by one-third. But this can be done only if a large part of the cotton crop is mechanized in preharvest and harvest work, if combining is increased substantially in the East and South, if corn pickers come into much greater use, and if hay-making methods are generally adopted that will eliminate pitching and hand loading of hay, and hand work at the barns and stacks and in the mows. It can be done only if many operations on small farms as well as on large farms are mechanized.

The least opportunity for eliminating crop hand work is for such crops as fruits, berries, nuts, vegetables, tobacco, and potatoes. Most of the crops in these groups must be handled carefully, and in some instances must be packaged in a way that will conserve their attractiveness and quality. Farm-maintenance work, amounting to about 15 percent of all farm labor, is mostly hand work. The hammer, saw, and trowel are principal tools in the erection and maintenance of farm buildings. Similar hand tools are used for cutting posts, building and repairing fences, and keeping machines and tools in order. These activities are not so susceptible to mechanization as are harvesting wheat, picking corn, harvesting hay, digging potatoes, milking cows, and pumping water.

Possibilities of reducing hand-hours by other means than mechanization are very great. Undoubtedly many of the 13 billion hand-hours are performed because of poor hand tools—a dull ax or saw, a worn-out hand tool, unsuitable hand tools for the jobs, etc. Poorly arranged buildings and feeding devices may double the time it takes to do the chores.

Recent farm-work simplification studies have indicated extensive possibilities in reducing hand-hours, and in doing the jobs with less effort, by scientific application of the laws of hand motion. The time required to do such hand jobs as setting plants, cutting seed potatoes, harvesting tobacco, harvesting celery, and tending livestock has been cut from 20 to 50 percent in conducted experiments. In one instance a Vermont dairy farmer, in cooperation with the Vermont Agricultural Experiment Station,[7] by making a series of inexpensive changes in and around his dairy barn, cut the time to tend 22 dairy cows by about 2 hours per day, and reduced the distance walked by 2 miles a day.

All hand jobs are not subject to such labor reductions. The time required to load hay of a given yield when a hay loader is used is governed by the speed with which the hay comes onto the wagon, but the time required to pick a bushel of tomatoes of a given yield is influenced by the waste motion of the picker. It seems probable that by far the greatest reduction in man-hour requirements so far has come about through increased use of improved machines. The 13 billion hand-hours in agriculture offer a large and profitable field for future savings through better arrangements of working facilities, better hand tools, and less waste motion in doing specific jobs.

CHANGES IN PATTERN OF MECHANIZATION

On January 1, 1910, the value of machinery on farms, including automobiles, tractors, and motortrucks amounted to more than 1.2 billions of dollars; on January 1, 1946 all machinery on farms was valued at about 6.3 billions of dollars, an increase of about 400 percent.

---

[7] Carter, R. M. Labor Saving Through Farm Job Analysis (Dairy Barn Chores). Vt. Agr. Expt. Sta. Bul. 503. June 1943.

PROGRESS OF FARM MECHANIZATION

This remarkable upward trend in inventory values of farm machinery was accentuated sharply during the first and second World War periods, and was slowed down considerably during the depression years of the 1930's (fig. 4).

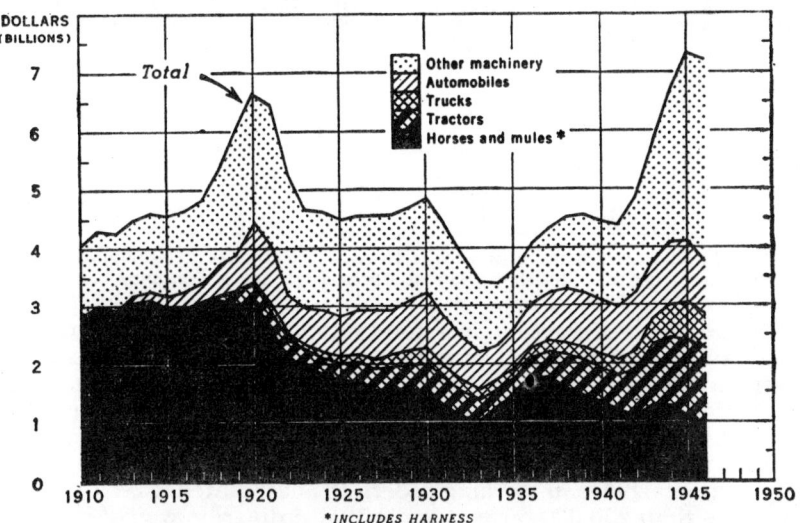

FIGURE 4—INVENTORY VALUES OF FARM HORSES AND MULES, AND FARM MACHINERY, JANUARY 1, UNITED STATES, 1910-46.
(VALUE IN CURRENT DOLLARS)

Increasing values of automobiles, tractors, and motortrucks have largely offset the declining total value of all horses and mules on farms.

Although price levels at various stages of the upward swing over the 35-year period influenced total values, by far the greatest change was caused by the increase in automobiles, tractors, and trucks which, as a group, rose in value from practically nothing in 1910 to a peak of 3 billion dollars on January 1, 1945. By January 1, 1930, farm automobiles had risen to 96 percent of the 1942 numbers—the highest number ever recorded for our country. The number of motortrucks on farms at the beginning of 1930 was large also—about 60 percent of the high number of January 1, 1946. Tractor numbers in 1930, on the other hand, were only 36 percent of the high number reported on January 1, 1946 (table 17).

Thus, all three—automobiles, tractors and trucks—were responsible for increasing machinery inventories during the earlier part of the period 1910-45. But of the three, farm tractors have been responsible for most of the increase in inventories since 1930. This is noteworthy when considered along with the declining numbers of farm horses and mules. It means that all three machines were responsible for horse-and-mule displacement in the earlier part of the period, but that tractors were very largely responsible for this displacement in the latter part of the period. At present, increases in tractors on farms measure fairly well the reduced need for horses and mules on farms.

TABLE 17.—*Numbers of tractors and other specified machines on farms, United States, January 1, 1910–46*

Year	Farm tractors	Farm motortrucks	Farm automobiles	Grain combines	Corn pickers	Milking machines
	Thousands	Thousands	Thousands	Thousands	Thousands	Thousands
1910	1	0	50	1	....	12
1920	246	139	2,146	4	10	55
1930	920	900	4,135	61	50	100
1940	1,545	1,047	4,144	190	110	175
1941	[1] 1,675	[1] 1,095	4,190	225	120	210
1942	1,890	1,160	4,285	275	130	255
1943	2,100	1,280	4,175	320	138	309
1944	2,210	1,370	4,120	345	146	345
1945	2,425	1,460	4,100	375	168	379
1946	2,585	1,550	4,100	415	200	450

[1] 1941–44 data are revised estimates of Bureau of Agricultural Economics, adjusted to preliminary census numbers; 1945 numbers are from preliminary census report.

The tremendous wartime increase in numbers of tractors and tractor labor-saving machines and motortrucks explains in large measure how fewer farmers with fewer hired and family workers did a much larger production job. Not until after the findings of the 1945 agricultural census became available did anyone fully realize the extent to which manufacturers, servicemen, farmers, and others had worked together to keep farm machines in working order. It is estimated that during the war more than 250,000 farm tractors that ordinarily would have been junked were kept going in this way as were many automobiles, trucks, combines, and other machines.

Many of these will probably be junked in the near future as new machines become available. But some of them may remain on farms for several years as a second or third tractor or machine; and some may find their way to small farms where the work load is light, and where the cost of new tractors and machines seems too high for the available volume of business.

When horse and mule values, including harness, are added to machinery values the total increase in farm animal and mechanical power and farm machines is much less pronounced than is the increase when animal power is omitted. For example, the total value of all machinery and horses and mules on farms on January 1, 1910 was about 4 billion dollars compared with a peak of 7.3 billion dollars on January 1, 1945, an increase of only 80 percent compared with the increase of 400 percent when horse and mule values are omitted from the totals. This large difference is explained by the fact that the inventory value of horses and mules and harness on January 1, 1945 was only 1.1 billion dollars compared with 2.8 billions at the beginning of 1910. Peak current inventory values before World War II were recorded in the postwar year 1920, and low inventory values occurred in the depression year 1934.

If allowances are made for changes in prices of farm horses and mules and for machinery, the inventory volume of total farm power and machinery seems to have been remarkably constant during the last 25 years, with the exception of the years immediately following World War I and the years of World War II when machinery purchases were very large, and the 1930's when machinery purchases were at a long-time low (fig. 5). During the first 8 years of the 37-year period, numbers of horses and mules, automobiles, and horse-drawn machines were increasing. This increase was continued well into the 1920's, except in the case of horses and mules, whose decline was more than offset by increasing numbers of tractors, tractor machines, and motortrucks. From 1910 to 1916, harvested crop acreages increased at an average rate of about 1 percent per year, and volume of farm machinery and power, exclusive of automobiles, at about the same rate. If automobiles are included, the average rate of increase was about 2 percent per year. After World War I large purchases of machinery, including automobiles, boosted the inventory volume of machinery and power on farms higher in relation to crop acreages than ever before. The same thing happened again in World War II period, and the upward swing in farm mechanization is still very strong.

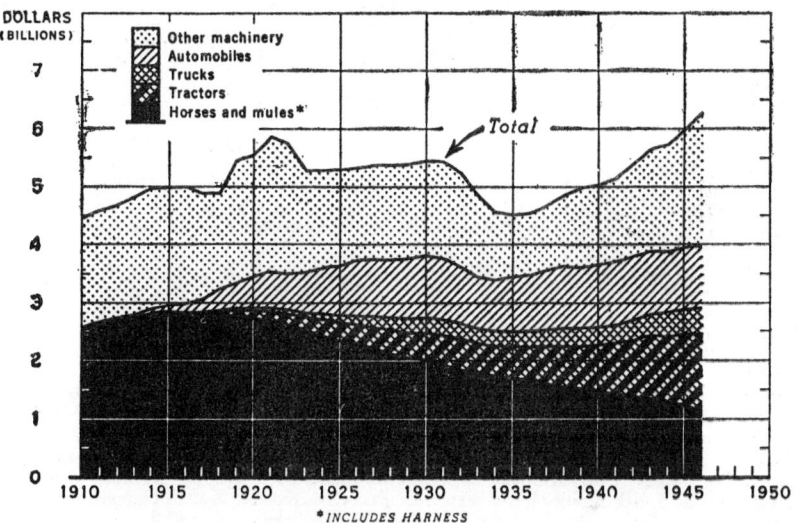

FIGURE 5—INVENTORY VALUES OF FARM HORSES AND MULES, AND FARM MACHINERY, JANUARY 1, UNITED STATES, 1910-46.
(VALUE IN 1935-39 AVERAGE DOLLARS)

When adjustments are made in the inventory values shown in figure 4 for changes in price levels, the inventory value, or volume, of all farm power, machinery, and equipment has varied remarkably little during most of the last 25 years. More automobiles, motortrucks, tractors and tractor machines, and equipment have made up for fewer horses and mules, and horse-drawn machines and equipment.

Displacement of Work Animals by Mechanical Power

During World War I and during the first half of the 1920's farm tractors were used almost exclusively for the heavier farm jobs, as plowing and disking, and for belt work. In the early days of their use a farmer who bought a tractor usually disposed of only 2 to 3 horses, and this practice was followed on the larger farms that had a good many horses. As the general-purpose type of tractor came into use, and as more and better tractor machines and equipment were made, more horses could be disposed of when a tractor was bought.

Displacement of animal power by farm tractors has not been uniform in all parts of the country according to a survey in 1943 of 10,022 tractor- and animal-operated farms (table 18). This study showed that the number of work animals displaced per farm tractor averaged only 1.8 in the Northeastern States compared with 3.4 in the Corn Belt, 5.3 in the Great Plains wheat and range country, and 10.6 on the farms reporting in the Pacific Coast States. The average displacement for all farms reporting was 4.4 head of workstock.

TABLE 18.—*Size of farms, cropland harvested, and numbers of farm tractors and horses and mules 3 years or more of age on farms reporting tractors and work stock on January 1, 1943* [1]

State group [2]	Size of farm in 1942	Cropland harvested per farm in 1942	Tractors per farm January 1, 1943	Work animals per farm January 1, 1943	Work animals displaced per farm tractor [3]
	Acres	*Acres*	*Number*	*Number*	*Number*
Northeast	197	98	1.20	2.7	1.8
Corn Belt	237	152	1.25	3.1	3.4
Lake States	219	132	1.19	3.1	2.3
Great Plains	755	340	1.33	4.4	5.3
Appalachian	292	113	1.09	3.8	2.2
Southeast	381	143	1.18	4.1	2.1
Delta States	600	309	1.80	10.5	5.8
Oklahoma-Texas	894	294	1.31	3.8	8.6
Mountain	1,709	265	1.28	5.5	7.3
Pacific	920	297	1.43	3.8	10.6
United States	522	202	1.27	3.8	4.4

[1] Based on 10,022 reports for tractor farms having work animals, distributed as follows: Northeast, 1,120; Corn Belt, 2,676; Lake States, 1,441; Great Plains, 1,752; Appalachian, 379; Southeast, 528; Delta, 214; Oklahoma-Texas, 542; Mountain States, 698; and Pacific, 672.

[2] Northeast includes Maine, New Hampshire, Vermont, Massachusetts, Rhode Island, Connecticut, New York, New Jersey, Pennsylvania, Delaware, and Maryland. Corn Belt includes Ohio, Indiana, Illinois, Iowa, and Missouri. Lake States include Michigan, Wisconsin, and Minnesota. Great Plains include North Dakota, South Dakota, Nebraska, and Kansas. Appalachian includes West Virginia, Kentucky, and Tennessee. Southeast includes Virginia, North Carolina, South Carolina, Georgia, Florida, and Alabama. Delta includes Mississippi, Arkansas, and Louisiana. Mountain includes Montana, Idaho, Wyoming, Colorado, Utah, Nevada, New Mexico, and Arizona. Pacific includes Washington, Oregon, and California.

[3] It was assumed that if there were no tractors the number of work stock needed in each State group would be the same per 100 acres of cropland as was reported for horse or mule farms. From the calculated number of needed work animals on tractor farms, the actual number of work stock on tractor farms was deducted and this figure divided by the average number of tractors gives the work animals displaced per tractor.

The average size of all tractors on farms in 1942 probably was around 15- or 16-drawbar horsepower (rated load basis). The average tractor of 15 or 16 horsepower therefore displaced only 4.4 horses and mules of work age. It is evident that tractors have a large surplus of power. Probably few farm jobs use this surplus. Consequently, it is not surprising that, at a given time, the inventory value of the average farm tractor is more nearly in proportion to the average value of the workstock displaced than to the drawbar power displaced, especially after farms became well supplied with automobiles and trucks. The 1935-39 average inventory value of tractors on farms was about $500 and the average value of the estimated 4.4 head of work animals displaced by a tractor in 1942 was about $450.

This apparent per tractor displacement of work animals on farms that have tractors does not represent the long-time disappearance of work horses and mules from farms as tractor numbers increased. For example, from 1919 to 1924 tractors on farms increased by 338 thousand but horses and mules of work age decreased by only 10 thousand. This is a decrease of only 0.03 of a work animal for each tractor added to farm inventories during the 5 years (table 19). In this postwar period of farm mechanization there were several million head of colts and young work animals on farms that had been produced because of prewar city and farm demand, and because of wartime civilian and military needs. Large colt crops previous to and during this postwar period resulted in a decline of only 10 thousand animals of work age whereas the number of horses and mules decreased by more than 3 million head. For many years during this transition period, when mechanical power was increasing rapidly, farms as a whole were apparently over-equipped with motive-power units.

TABLE 19.—*Disappearance of horses and mules from farms, and increase in farm tractors and automobiles and trucks, by periods, United States, 1919–44*

Period of change	Increase in tractor numbers on farms	Decrease in number of horses and mules on farms		Disappearance from farms of horses and mules, per tractor		Increase in numbers of all automobiles and motor trucks on farms
		Animals of all ages	Animals 3 years old and older	Animals of all ages	Animals 3 years old and older	
	Thousands	*Thousands*	*Thousands*	Number	Number	*Thousands*
Change from January 1:						
1919 to 1924	338	3,205	10	9.48	0.03	1,496
1924 to 1929	331	3,541	2,610	10.70	7.89	1,443
1929 to 1934	189	2,747	2,275	14.53	12.04	[1] 536
1934 to 1939	429	2,205	3,035	5.14	7.07	776
1939 to 1944 [2]	765	2,179	1,496	2.85	1.96	440
1919 to 1939	1,287	11,698	7,930	9.09	6.16	3,179
1919 to 1944	2,052	13,877	9,426	6.76	4.59	3,619

[1] A decrease.
[2] 1939–44 computations made from revised estimates of Bureau of Agricultural Economics.

During the next 5 years 1924–29, tractor numbers increased about the same as during 1919–24 but workstock numbers decreased by 2.6 million head. The decrease in workstock numbers for each tractor added to farm inventories during this 5-year period was about 7.9. In the following 5-year depression period, 1929–34, workstock continued to decrease rapidly but tractors increased rather slowly. More than 12 head of workstock disappeared from farms for each tractor added.

During the war years, 1939–44, the increase in tractors was almost 80 percent greater than in the previous 5 years. Even though numbers of workstock decreased by about 1.5 million head, therefore, tractor numbers increased so much more rapidly that workstock displacement amounted to only about 2 head per tractor added to farm numbers during these war years. Although these comparisons are in terms of average disappearance per tractor, it must be remembered that increasing numbers of automobiles and trucks helped to effect this displacement, especially in the earlier years (table 19).

Figure 6 shows the rapid increase that has taken place in the number of tractors on farms, and the decrease in number of farm horses and mules, since World War I. The first large increase in tractor numbers came after the war's end, when agricultural and industrial production for peacetime consumption were at high levels. Development of the general-purpose type of tractor, and its widespread adoption by farmers in the late 1920's was chiefly responsible for maintenance of the rapid upward trend in the number of tractors on farms. In the late 1930's rubber-tired general-purpose tractors came on the markets. These were bought freely by farmers, and the increase in tractor numbers, which had slowed down during the depression years, was resumed. The present large postwar demand for farm tractors and tractor equipment points to a continuation of the strong upward swing in farm mechanization. Horses and mules of work age will continue to disappear, unless the downward trend in colt production is halted.

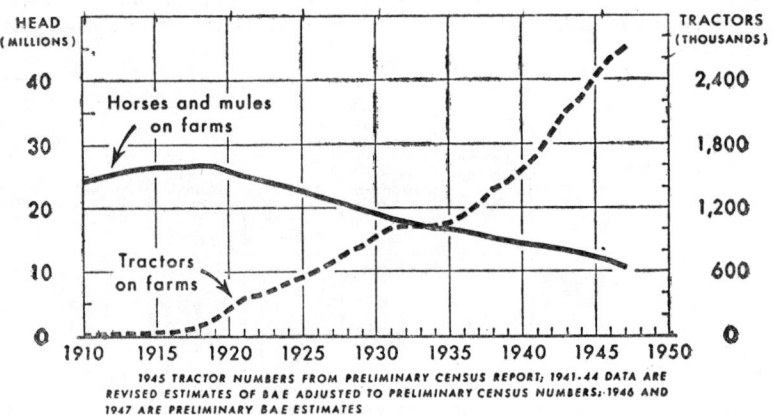

FIGURE 6—HORSES AND MULES, AND TRACTORS ON FARMS, JANUARY 1, UNITED STATES, 1910-47.

Horses and mules, including colts, reached peak numbers in 1918, and have declined since. Colts are now insufficient to maintain the numbers of horses and mules of working age. Tractors had become basic equipment on many farms by 1930, and were added rapidly thereafter as farmers were able to buy a tractor and tractor equipment.

Tractors and Tractor Equipment

The starting place in saving crop labor by mechanization is the farm tractor. By itself, it saves no labor—it is merely the power unit for machines and tools that, combined with more power and a higher rate of travel, do farm jobs faster. Two important machines that have been developed along with tractors are grain combines and mechanical corn pickers. In 1910 there were 1,000 farm tractors and 1,000 grain combines. Most of the latter were being operated with horses and mules. In 1946, there were about 2.6 million farm tractors and 415,000 combines, practically all of which were operated with tractors. Mechanical corn pickers increased from 10,000 in 1920 to 200,000 in 1946. Milking machines helped to save labor in the dairy barns. In 1946, there were 450,000 milking-machine installations compared with only 12,000 installations in 1910 (table 17).

Similar data for several other tractor- and horse-operated machines are available for each year during 1942–45. There were more of most tractor-operated machines and fewer of each horse-operated machine on farms in 1945 than in 1942 (table 20). Windrow pick-up balers increased by 67 percent. Peanut pickers, corn pickers, grain combines, and tractor-operated cultivators, mowing machines, and row-crop planters increased by about 25 percent on the average. Tractor-operated plows, listers, and middlebusters of all kinds and sizes increased 10 percent from January 1, 1942 to January 1, 1945. On the other hand, the general rule was for horse-operated machines and tools to decrease in numbers by about 5 to 10 percent, and horse-operated stationary hay balers decreased by 12 percent.

TABLE 20.—*Estimated number of specified machines and implements on farms on January 1, 1942 and 1945, United States totals* [1]

Machine or implement	1942	1945	Percentage 1945 is of 1942
	Thousands	*Thousands*	*Percent*
Manure spreaders—mostly horse-drawn	1,159.7	1,128.5	97
Moldboard plows—tractor-drawn	1,468.4	1,616.4	110
Moldboard plows—horse-drawn	7,670.8	7,239.6	94
Disk plows—tractor-drawn	167.2	176.3	110
Disk plows—horse-drawn	83.1	73.9	89
One-way disk plows or tillers—tractor-drawn	160.9	168.5	105
Listers and middlebusters—tractor-drawn	292.2	313.1	107
Listers and middlebusters—horse-drawn	1,277.3	1,217.0	95
Disk harrows—tractor-drawn	1,189.7	1,262.6	106
Disk harrows—horse-drawn	1,337.3	1,244.6	93
Row-crop planters—tractor-drawn	205.2	255.6	124
Row-crop planters—horse-drawn	3,419.8	3,238.1	95
Row-crop cultivators—tractor-drawn	889.3	1,171.2	131
Row-crop cultivators—horse-drawn	7,072.4	6,764.0	96
Grain drills—tractor-drawn	428.4	420.6	98
Grain drills—horse-drawn	1,277.9	1,220.8	96
Grain binders—tractor-drawn	365.6	337.1	92
Grain binders—horse-drawn	1,015.8	909.3	91
Grain combines—tractor-drawn	275.0	375.0	136
Grain separators—threshers	167.9	152.8	91
Mowing machines—tractor-drawn or mounted	313.7	395.2	126
Mowing machines—horse-drawn	2,571.1	2,423.5	94
Rakes, sulky or dump	2,164.6	2,084.3	93
Rakes, side delivery	715.6	739.1	103
Balers, windrow pick-up—tractor-drawn	25.1	41.9	167
Balers, stationary—power-operated	66.2	65.3	99
Balers, stationary—horse-operated	61.6	54.0	88
Row-binders—tractor-drawn	82.3	86.4	105
Row-binders—horse-drawn	526.8	491.8	93
Corn pickers—tractor-drawn	130.0	168.1	129
Peanut pickers	8.5	11.2	132
Cream separators	1,750.5	1,751.0	100
Milking machines	254.7	379.3	149

[1] Brodell, A. P., and Cooper, M. R., Number and Duty of Principal Farm Machines, Bureau of Agricultural Economics, F.M. 46, Nov. 1944. (Processed.)

At the beginning of 1942 farmers had more than 11 million plows, listers, and middlebusters, of all kinds and sizes, or an average of about 2 per farm. By January 1945 the total had decreased by about 315 thousand, but the decrease was in the less effective horse-drawn plows and listers, and there were more tractor-drawn plows and listers. The change undoubtedly meant an increase in plow and lister capacity from 1942 to 1945. Total numbers of other machines, like row-crop planters and cultivators, and mowing machines, decreased from 1942 to 1945, but those operated with tractor power increased so that here again the work capacity of the equipment was increased.

Numbers of both horse-drawn and tractor-drawn grain binders decreased during 1942–45. But these were being replaced by increasing numbers of a newer and more efficient type of machine—the combine harvester-thresher. Numbers of stationary grain separators decreased also. Mechanical corn pickers increased during the period by almost 30 percent. These additional machines probably now pick 250 to 275 million bushels of corn that were picked by hand in 1942. Pick-up hay balers probably now bale more than double the quantity of hay and straw that was baled in this way in 1942.

Farmers of the United States, in general, own most of their farm machines and tools, especially the smaller and less expensive ones, such as plows, cultivators, and planters. Machines like tractors, trucks, hay balers, corn pickers, and grain threshers are expensive for many farmers. They are more suitable for joint ownership by small producers and for outside custom and exchange work. During World War II there was a decided increase in the consolidation of farms in some areas. Some of these were partnership consolidations of two or more units owned by close relatives. These consolidations included the farm machines and livestock as well as the land. In some instances of this nature the size of the farm was increased to take advantage of available machinery and labor, but the number of farm families deriving income from the farms remained the same.

Present inventories of farm machines include machines on many farms that are not fully used. Timeliness of use of some machines is so important that the farmer prefers to own the machine even though its annual use is very restricted. Some of the machines and tools in farm inventories are large horse-drawn machines for which there is no great use in the locality. Then, there are modern machines for which there is only limited outside work.

The milking machine, ordinarily used twice during nearly every day in the year, was used more in 1941 than any other of the machines listed in table 21, or an average of 684 hours per year. Farm tractors were used an average of about 500 hours per year, and cream separators about 140 hours. Planters, binders, mowing machines, and drills were usually used well under 100 hours per year.

Undoubtedly, the general tendency during the war was to increase the average annual use of labor-saving machines. The range in use of machines on individual farms is very wide. Some of the smaller and less expensive tools are sometimes used no more than a full day each year—but they may be very essential at just the right time. Tractor-drawn equipment generally is used on larger farms than horse-drawn machines, and is used more hours and does more work per year than horse-drawn equipment. For example, 2-row tractor-drawn row-crop planters were used to plant an average of 131 acres in 1941, but 2-row, horse-drawn planters averaged only 46 acres. Tractor-drawn or mounted mowing machines of all sizes cut an average of 154 acres in 1941, compared with an average of 54 acres per horse-drawn mower of all sizes.

TABLE 21.—*Extent of use and work done in 1941 for indicated kinds, and sizes of machines* [1]

Kind of machine	Size of machine and power used	Time machine was used in 1941		Average amount of work done with machine in 1941
		Usual range	Average	
		Hours	*Hours*	*Acres*
Row-crop planters	2-row, tractor-drawn	20–250	76	131
	4-row, tractor-drawn	20–250	80	262
	2-row, or more, horse-drawn	10–250	40	46
	1-horse	5–300	50	28
Row-crop binders	1-row, tractor-drawn	10–250	51	38
	2-row, tractor-drawn	20–350	91	130
	Various sizes, horse-drawn	10–250	34	20
Mowing machines	Various sizes, tractor-drawn or mounted	20–500	78	154
	Various sizes, horse-drawn	10–400	63	54
Grain drills	Various sizes, tractor-drawn	20–500	79	201
	Various sizes, horse-drawn	10–350	44	44
Grain binders	Various sizes, tractor-drawn	20–350	55	100
	Various sizes, horse-drawn	10–225	34	37
Grain combines	6 feet and less, tractor-drawn	25–400	110	126
	Over 6 feet and under 10, tractor-drawn	25–400	126	207
	10 feet and over, tractor-drawn	25–400	143	400
				Tons
Manure spreader	Various sizes, horse- and tractor-drawn	25–1,000	142	177
Farm tractors [2]	Various sizes and types	100–1,400	493	....
Milking machines	Various sizes	200–2,000	684	....
Cream separators	Various sizes	50–800	139	....

[1] Brodell, A. P., and Birkhead, James W., Work Performed With Principal Farm Machines. Bureau of Agricultural Economics, F.M. 42, May, 1943. (Processed.)
[2] Brodell, A. P., and Cooper, M. R., Fuel Consumed and Work Performed by Farm Tractors. Bureau of Agricultural Economics, F.M. 32, March, 1942. (Processed.)

Not only are tractor-drawn machines used more hours per year than are horse-drawn machines, but principally because of larger machines and higher rates of travel, they do much more work per hour. Table 22 gives several examples of an average day's work with different sizes of machines, operated with specified units of animal or tractor power. These illustrate a compelling influence in the change in pattern of farm power and machines over the last 20 to 30 years. One man with a 2-bottom, 14-inch moldboard plow drawn by 5 horses will plow 4 acres in a 10-hour day, but with a tractor plow of the same size and a 15-horsepower tractor he will plow 8 acres in the same time—and if necessary he can keep going for 12, 14, or more hours per day, whereas the horses after 10 hours in the field must rest until the next day.

TABLE 22.—*Daily duty of machines of specified kinds and sizes*

Operation	Kind and size of implement [1]	Power used [2]	Average acres covered per 10-hour day
Plowing	Moldboard, walking, 8-inch	1 horse	1.0
	Moldboard, walking, 14-inch	2 horses	2.0
	Moldboard, 2-bottom, 14-inch	5 horses	4.0
	Moldboard, 2-bottom, 14-inch	15-h.p. tractor	8.0
	Disk plow, 50-inch	20-h.p. tractor	13.0
	Disk plow, 10-foot, vertical, one way	20-h.p. tractor	28.0
Disking	Single disk, 8-foot, once over	4 horses	15.0
	Single disk, 20-foot, once over	20-h.p. tractor	60.0
Harrowing	Spike-tooth, 10-foot, once over	2 horses	15.0
	Spike-tooth, 20-foot, once over	4 horses	30.0
	Spike-tooth, 24-foot, once over	15-h.p. tractor	70.0
	Spike-tooth, 32-foot, once over	20-h.p. tractor	90.0
Cultivating Corn or cotton	½-row, walking (2 times to row)	1 horse	2.5
	1-row, riding (1 time to row)	2 horses	7.0
	2-row, riding (1 time to 2 rows)	3 horses	12.0
	2-row, riding (1 time to 2 rows)	15-h.p. tractor	20.0
	4-row, riding (1 time to 4 rows)	15-h.p. tractor	35.0
Cutting corn	Binder, 1-row (corn not shocked)	3 horses	6.5
	By hand (corn shocked)	Hand	1.2
Picking corn	Mechanical picker, 1-row	15-h.p. tractor	7.0
	Mechanical picker, 2-row	20-h.p. tractor	12.0
	By hand	Hand	1.5
Mowing hay	Mower, 5-foot	2 horses	8.0
	Mower, 7-foot	15-h.p. tractor	20.0
Drilling grain	Disk drill, 7-foot	3 horses	12.0
	Disk drill, 10-foot	6 horses	18.0
	Disk drill, 10-foot	20-h.p. tractor	25.0
	Disk drill, 20-foot	20-h.p. tractor	50.0
Harvesting grain	Binder, 6-foot	3 horses	9.0
	Binder, 8-foot	4 horses	14.0
	Binder, 8-foot	15-h.p. tractor	20.0
	Combine, 5-foot	15-h.p. tractor	11.0
	Combine, 10-foot	20-h.p. tractor	22.0
	Combine, 16-foot	20-h.p. tractor	30.0
Milking cows [3]	Milking machines, 2 unit (one milking)	Electric motor	[4] 3–5
	Milking by hand (one milking)	Hand	[4] 8–10

[1] Size represents working width.
[2] For modern tractors equipped with rubber tires.
[3] Minutes per cow include time for caring for machine and milking utensils.
[4] Minutes per cow.

For obvious reasons these larger, faster, and more expensive power and power-operated machines were first used almost exclusively on the larger farms. Gradually, machines were manufactured more and more in line with the needs of farmers who had commercial family-sized farms and, to some extent, to the needs of less than full-time farmers. On large farms in the Great Plains one man with a 10-foot one-way plow and a modern 20-horsepower tractor will plow an average of 28 acres in a day, which is more land than many farmers on small farms will have available for plowing in an entire season. A 10-foot spike-tooth harrow drawn by 2 horses will cover, once over, about 15 acres in a day. A few days of work a year with this size of harrow will do all the harrowing there is to be done on many small farms. But the large operator with 1,500 or 2,000 acres of semi-arid cropland must

PROGRESS OF FARM MECHANIZATION

do the bigger job with bigger equipment. With a 32-foot spike-tooth harrow and a 20-horsepower modern tractor he will harrow 90 acres in a day—a full week of work with the smaller horse-drawn harrow.

Generally, large machines are associated with large acreages, large fields, and extensive types of farming, and small machines with small acreages, small fields, rough lands, and some of the more intensive types of agriculture. The size of the job to be done and the conditions under which the work is to be done help to determine the selections to be made from the various kinds and sizes of machines. Our farmers have a wide choice in kinds and sizes of farm machines and tools, and may make their selections according to the needs of their particular situations. Thus, changes in the pattern of farm mechanization in the different localities and regions may be influenced more often by type of farming and available physical and financial resources than by other factors.

Effect on Timeliness of Operation

No precise measure is available of the effect of mechanization on production because of timeliness in doing farm operations. Yet much evidence points to greater production and higher quality of product because of better timing of operations, made possible mainly by up-to-date power units and machines. Trading centers now in many cases are several hours closer to the farm than in days of poor roads and slow teams. A broken part for the hay loader may be brought from town and installed and the haying operation resumed in a few hours, whereas formerly the storing of hay under cover might be postponed until the next day. And a modern machine shop on the farm might reduce the lost time considerably, especially in the latter case. More acres per hour with modern power machines—and more hours per day when necessary—help to get the critical jobs done right and on time. Facilities for better and more timely grading, cooling, and storing, insure better products in the markets.

But the real advantages often come in seasons of adverse working conditions. These are the critical times when the difference in equipment and power means the difference between little or no crop and a large crop.

This is how one farmer illustrated the point. In a dry, hot fall he bought a tractor and equipment because he could not plow the hard ground in time for seeding wheat with the power he had available. He prepared the ground and seeded 85 acres of wheat with the new outfit, compared with a possible 10 or 15 acres he could have done with the old horse-drawn tools. He said that the difference in costs between animal and tractor power was unimportant under the circumstances—he had more than 2,000 bushels of wheat to sell the following July instead of a possible 2 or 3 hundred bushels, and his farming scheme was not interrupted. Size of business was important in the case cited. Small farmers may not be able to switch so readily when the purchase of a tractor and equipment is involved, but probably many can meet such emergencies by hiring the work done by custom operators.

An excellent illustration of the contribution of mechanical power to timeliness of operations is the experience of Corn Belt farmers in the very wet spring seasons of 1943, 1944, and 1945. In the corn-planting month of May 1943, for example, rainfall in Illinois was 8.75 inches, or more than double the usual precipitation. The fields were so wet that only 15 percent of the corn crop was planted by June 1, which is generally about the outside date for planting in the State. By utilizing all available mechanical power and equipment the corn lands were prepared, and the remaining 85 percent of the crop was planted in the first 2 weeks after the rains stopped and the land became workable. Some of the machines came from off-farm places, and many of them were operated on a 24-hour schedule. As the season advanced, earlier maturing strains of seed were brought in from farther north, and by June 15 the basic job had been completed. The corn crop that year was the largest Illinois had ever had.

If tractor power had not been available large acreages of corn could not have been planted in time to get a crop. Other large acreages would have been in poor condition, and many late plantings would have produced little solid corn. The speeding up of work in Illinois with power equipment was brought about in two ways—by working more continuous hours and by doing more work per hour. With a tractor and the tractor equipment commonly used in Illinois, 3 acres can be prepared and planted to corn in the same time required to prepare and plant 1 acre with animal power and equipment. When the tractor is put on a 24-hour schedule (an impossibility with work animals) 7 acres can be prepared and planted as compared with 1 acre with animal power.

The example of the tractor's accomplishment in Illinois probably could be duplicated on a smaller scale somewhere in any season. The cumulative effect of the numerous cases of "late seasons" on production must be very large in some years. It is entirely possible that the great amount of work that can be done in a short time with mechanical farm equipment has largely eliminated low production caused by the narrowing of planting seasons by rains and storms. It is possible also that weather hazards in the cultivating and harvesting seasons are not so great as when the work was done by time-consuming methods. The working facilities, provided by tractor farming, combined with the greater range in the number of days required to mature crops from various strains and varieties of seeds, and a more widespread knowledge of crop responses to various applications of fertilizer, have brought about a higher average yield of our major crops than previously was possible. This does not mean, however, that all of the fluctuations in yields and production have been eliminated. Damage to growing crops because of droughts, storms, freezes, diseases, insects, etc. cannot be completely eliminated.

Regional Changes in Mechanization

Beginning on page 12 the idea was developed that high production per worker is generally found in those areas in which soils are productive and the topography and type of agriculture are suitable for labor-saving machines. It is in these areas that changes in mechanization have been most pronounced. These changes are fairly well indicated by changes in tractor numbers since 1920 (fig. 7). In 1920, tractors were most common on the grain and livestock farms of the Great Plains and Corn Belt. Local areas in California and the Pacific

Northwest were relatively well supplied with farm tractors. Heavy concentration of farm tractors are now found in the Corn Belt, the wheat areas, the fertile fruit and crop areas in the East and in the irrigated valleys of the West. A large proportion of the tractors on Southern farms are in the concentrated rice, sugarcane, and tree fruit and nut areas and in some cotton areas.

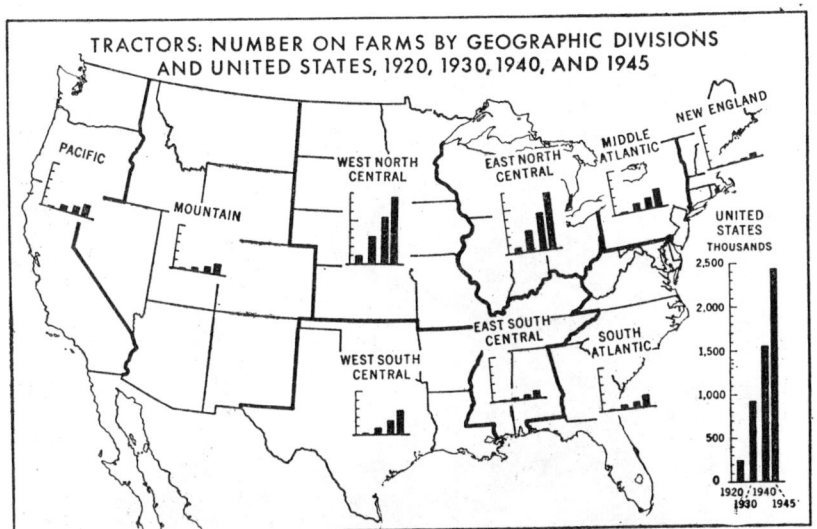

FIGURE 7—INCREASE IN TRACTOR NUMBERS HAS BEEN GREATEST ON FARMS IN THE CENTRAL AND SOUTHWESTERN PARTS OF THE UNITED STATES. SOUTHEASTERN FARMS REMAIN RELATIVELY UNDER-MECHANIZED.

The pattern of tractor distribution coincides closely with the pattern of volume of work in land preparation and cultivation per worker. More than 325,000,000 acres in the United States are seeded to crops each year, and most of this land is broken before being seeded (plowed, listed, or bedded). In 1939 about 55 percent of this breaking was done with tractor power, and the practice has increased decidedly since then. Disking and harrowing also require considerable power. Frequently, the optimum seasons for preparing land are short. In 1939 about 57 percent of the disking and 43 percent of the harrowing was done with tractor-drawn implements. The regional pattern for tractor disking and harrowing follows closely the regional pattern for land breaking with tractor power.

In 1939, 70 percent or more of large areas in the Corn Belt and in the western half of the United States were broken with tractor power (fig. 8). In many other important farming areas in the western half of the country and in the northern part of the eastern half of the country, tractor power was used for doing 50 to 70 percent of the land breaking. In large areas of the South, where many fields are small and many workers are needed for chopping and picking cotton, a small percentage of the land breaking was done with tractor power in 1939. In large southern areas as much as 90 percent of the breaking was done with animal power.

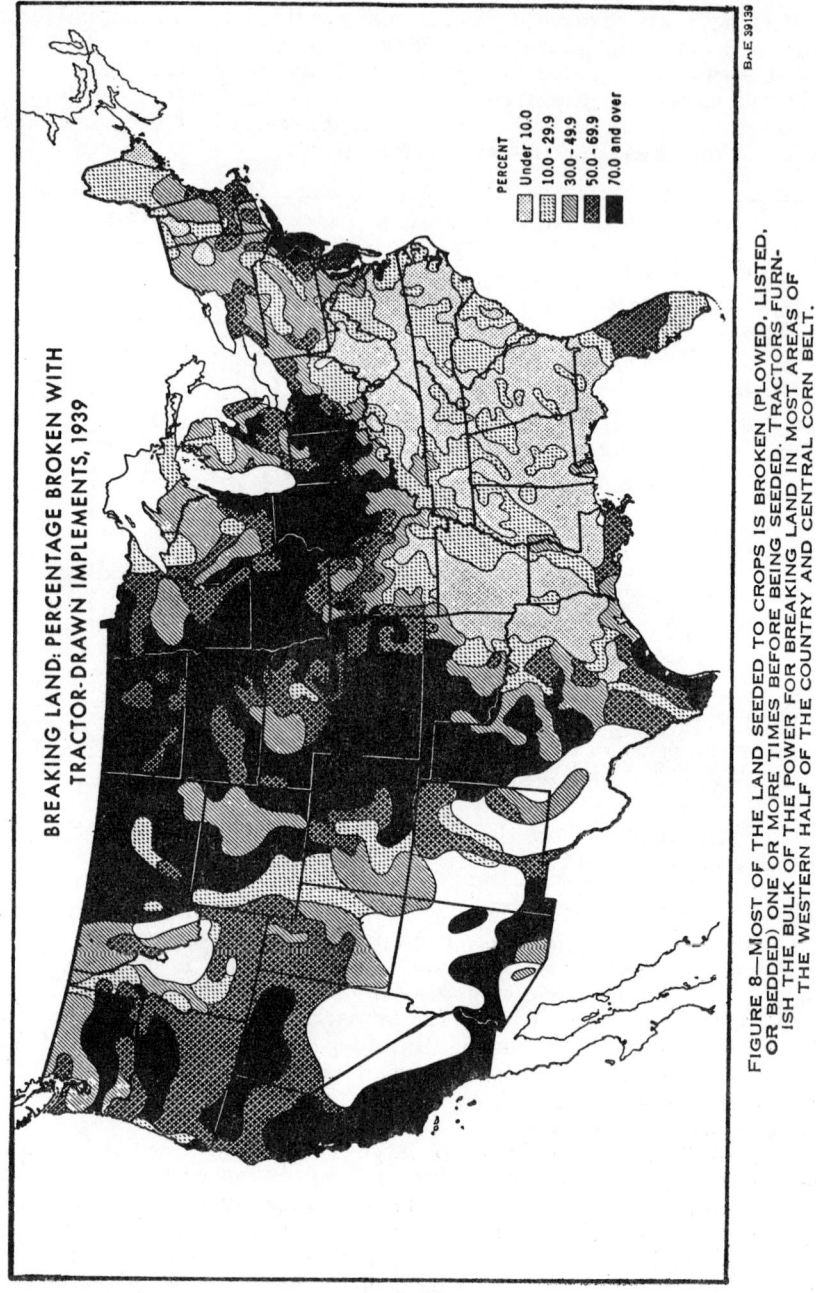

FIGURE 8.—MOST OF THE LAND SEEDED TO CROPS IS BROKEN (PLOWED, LISTED, OR BEDDED) ONE OR MORE TIMES BEFORE BEING SEEDED. TRACTORS FURNISH THE BULK OF THE POWER FOR BREAKING LAND IN MOST AREAS OF THE WESTERN HALF OF THE COUNTRY AND CENTRAL CORN BELT.

There is at present a large and growing unfilled demand in extensive areas of the South for farm tractors and tractor tools. A few mechanical cotton pickers are being used and plans call for the manufacture of cotton pickers in considerable volume. Many informed persons believe that mechanization of the cotton harvest will be a long, drawn-out process. However, because of increased cotton yields per acre more of the South's cropland is now being used for grain and forage, both of which are suitable for mechanization.

In the northeastern part of the United States farmers had more tractors per 100 acres of cropland in 1945 than was true in any other geographic division, including the rich, level north central divisions. Farmers in the East South Central States had the fewest tractors per 100 acres of cropland. The range from high to low, by geographic divisions, was 1.46 tractors per 100 acres of cropland in the New England States to only 0.34 of a tractor in the East South Central division (table 23). These differences do not necessarily indicate differences in extent of optimum mechanization in the divisions at a given time. Greater tractorization of Southern farms has had to wait until progress was made in other lines of mechanization. The relatively large number of tractors per 100 acres of cropland in the East does not necessarily mean waste of tractor power. The nature of the topography and the variety of crops mean more hours per acre of crops than are required in midwestern areas, whether the work is done with tractor or with animal power. The hours saved by using a tractor on crops in the East are especially valuable in caring for livestock on many poultry and dairy farms, even though the acreages are not large. Feed grinding, silo filling, and manure hauling are suitable for tractor use, but are not very important on noncommercial livestock farms.

TABLE 23.—*Number of tractors per 100 acres of total cropland,[1] by geographic divisions, 1920, 1930 and for the years 1940–45*

Geographic division	1920	1930	1940	1941	1942	1943	1944	1945
New England.......	0.05	0.36	0.76	0.88	0.97	1.15	1.28	1.46
Middle Atlantic.....	.08	.56	.91	.99	1.07	1.23	1.35	1.42
East North Central..	.09	.43	.77	.82	.92	.99	1.01	1.07
West North Central.	.07	.22	.40	.42	.48	.52	.51	.55
South Atlantic......	.03	.17	.21	.25	.28	.35	.39	.48
East South Central..	.02	.09	.15	.18	.20	.25	.27	.34
West South Central..	.04	.12	.29	.31	.37	.42	.44	.49
Mountain..........	.08	.17	.28	.31	.34	.38	.38	.42
Pacific............	.11	.34	.49	.53	.57	.62	.66	.71
United States.	.06	.24	.42	.46	.52	.57	.59	.64

[1] Includes acreages from which one or more crops were harvested, and acreages of crop failure and summer fallow.

Eastern farmers with their smaller acreages of cropland, have relatively high investments in farm machinery and power. In 1945 the total value of machinery and power (in terms of 1935–39 average dollars) was more than $38 per acre of cropland in the New England and Middle Atlantic States, compared with less than $12 per acre in the West North Central States (table 24). Even with the high proportion of relatively low-priced 1- and 2-mule tools in the South Atlantic and East South Central States, values of machinery and power were about $19 and $17 per acre of cropland, respectively. In only 1 of the 9 geographic divisions were values per acre, at 1935–39 prices, less in 1945 than in 1910. In 2 divisions total value of machinery and power

46 MISC. PUBLICATION 630, U. S. DEPT. OF AGRICULTURE

TABLE 24.—*Value of horses and mules, tractors, motortrucks, automobiles, and other farm machinery, per acre of cropland, by geographic divisions, 1910, 1920, 1930, 1940–45*

[Values at 1935–39 average prices]

Geographic division	1910	1920	1930	1940	1941	1942	1943	1944	1945
	Dollars	*Dollars*	*Dollars*	*Dollars*	*Dollars*	*Dollars*	*Dollars*	*Dollars*	*Dollars*
New England	27.90	27.64	33.11	29.27	30.00	31.25	33.49	34.02	38.75
Middle Atlantic	24.16	26.24	33.16	30.70	31.63	33.31	36.00	34.47	38.30
East North Central	14.49	16.75	16.90	18.99	19.20	19.84	20.42	20.22	20.62
West North Central	10.11	12.00	11.35	10.09	10.39	10.96	11.01	11.21	11.80
South Atlantic	14.05	17.08	16.08	15.60	16.20	16.42	16.58	17.21	18.72
East South Central	13.38	15.61	14.68	14.55	14.85	15.04	15.14	15.90	16.77
West South Central	11.52	11.12	10.77	10.13	10.52	11.23	11.63	11.68	13.10
Mountain	15.10	12.88	11.71	10.56	10.44	11.03	11.30	11.61	12.80
Pacific	12.22	16.02	16.51	16.64	17.51	17.97	18.82	18.89	20.55
United States	13.04	14.52	13.99	13.56	13.92	14.54	15.18	15.05	16.10

per acre, at 1935–39 prices, was slightly less in 1945 than in 1920, and change in the composition was noticeable in all divisions (see table 44 in the appendix).

The general pattern of change in the component parts of machinery and power per acre of cropland has been the same in each geographic division. Less investment in horses and mules, and more investment in tractors, trucks, and automobiles has been the general tendency. The rate of change, however, has been different in the different divisions. The relative importance per acre of horse and mule power (in terms of 1935–39 values) has declined less rapidly in the South Atlantic and East South Central Divisions than in other divisions, and the importance of tractors, trucks, automobiles, and other farm machinery has increased less rapidly than in most divisions. In 1945, for example, the 1935–39 dollar values of horses and mules per acre in the South Atlantic and East South Central States was only 30 percent less than in 1920; whereas in the other geographic divisions comparable decreases between 1920 and 1945 ranged from 44 to 69 percent (table 25).

TABLE 25.—*Change from 1920 to 1945 in the value at 1935–39 average prices of horses and mules, farm machinery, and total power and machinery, per acre of cropland, by geographic divisions*

Geographic division	Decrease from 1920 to 1945 in number of farm horses and mules, per acre	Increase from 1920 to 1945 in volume of farm machinery, including tractors, trucks, automobiles, and other machinery, per acre	Change from 1920 to 1945 in volume of all farm power and machinery, per acre
	Percent	*Percent*	*Percent*
New England	54	103	40
Middle Atlantic	44	101	46
East North Central	60	92	23
West North Central	62	47	−2
South Atlantic	30	66	10
East South Central	30	77	7
West South Central	50	110	18
Mountain	64	60	−1
Pacific	69	85	28
United States	54	71	11

For the United States as a whole, the 25-year increase in the 1935–39 dollar values per acre of machinery and motor vehicles more than offset the decrease in value per acre of horses and mules. Increases in farm machinery and power per acre were especially large during World War II when farmers bought so many farm tractors, motor trucks, and other labor-saving machines.

There are several reasons for the differences in value of machinery and power per acre of cropland. Some of these have been indicated. Within the same geographic region, the type and the size of farm business undoubtedly are influential factors. Farmers who have small businesses and low incomes cannot afford to buy expensive machines. Even the cheaper machines and tools may run the total investment per acre rather high if the crop acreage is small. This is not necessarily undesirable if the value per acre of farm products is large. Farmers who have larger acreages and low returns per acre may have difficulty in paying for their machinery even though the investment per acre is relatively low.

In 1939 about 40 percent of the total value of farm products was produced on farms having horses or mules but no tractors, 50 percent on farms having tractors alone or in combination with horses and mules, and 10 percent on farms having no horses or mules or tractors (table 26). Equipping a farm with power and machinery is an important problem for the individual farmer. But many farms are so small that, purely from the standpoint of efficient national food production, it makes little difference how well or how poorly they are equipped. In 1939, for example, each of 1,145,000 farms produced less than $250 worth of farm products, including the value of commodities consumed in the farm homes. These farms represented around 19 percent of the 6 million farms in the United States, but they produced only 2 percent of the value of all farm products. Farms in the next size-group shown in table 26 were low-producing farms also. It contained more than 1,692,000 farms that produced products valued at between $250 and $599 per farm. Thus, in 1939 there were more than 2,837,000 farms that produced less than $600 worth per farm. These 2.8 million farms made up almost one-half of the total number of farms in the United States, but they contributed only 11 percent of the total value of farm production.

TABLE 26.—*Farms reporting products valued at various amounts in 1939, and percentage of United States total value of products produced on farms with indicated types of power* [1]

Value of product group	Farms reporting value of products			Percentage of total U. S. products produced on farms reporting—				
	Farms	Percentage of all farms	Percentage of farms having tractors	Work animals but no tractors	Tractors and work animals	Tractors only	No tractors or work stock	All farms in value group
Dollars	*Number*	*Percent*	*Percent*	*Percent*	*Percent*	*Percent*	*Percent*	*Percent*
Under 250	1,145,005	19.2	4.7	0.9	([2])	0.1	1.0	2.0
250 to 599	1,692,245	28.4	7.7	5.8	0.5	0.2	2.4	8.9
600 to 999	1,053,575	17.6	16.4	7.1	1.3	0.5	1.6	10.5
1,000 to 1,999	1,124,998	18.8	36.7	11.0	6.5	1.2	1.6	20.3
2,000 to 3,999	639,993	10.7	61.3	7.4	12.4	1.6	1.1	22.5
4,000 to 9,999	254,626	4.3	75.7	3.5	12.4	1.8	1.0	18.7
10,000 and over	58,313	1.0	72.8	3.3	9.9	2.5	1.4	17.1
All groups	5,968,755	100.0	23.4	39.0	43.0	7.9	10.1	100.0

[1] Adapted from technical monograph, Analysis of Special Farm Characteristics for Farms Classified by Total Value of Products, United States Department of Commerce and United States Department of Agriculture, 1943. Value of product is for 1939 and numbers of farms, tractors, horses, and mules for April 1, 1940.

[2] Less than 1/10 of 1 percent.

Many of these low-producing farms are part-time or self-sufficing places, and economical production often is a matter of finding ways of increasing production rather than of using more labor-saving machines. For example, about three-fourths of the farms reporting products valued at less than $250 per farm were classified as subsistence farms, or farms on which the major source of income was from products used by the farm household. In the next group, composed of farms reporting products valued at $250 to $599, about one-half were classified as subsistence farms.

In the lowest value group (under $250) less than 5 farmers in each 100 reported having tractors in 1940, and about 53 in each hundred reported having no tractors, horses, or mules (table 27). The next size group, composed of farms producing farm products valued at $250 to $599, reported more tractor and animal power, but even in this group less than 8 farmers in each 100 owned tractors, and almost 30 percent reported no tractors, horses, or mules.

TABLE 27.—*Percentage of farms reporting tractors, work animals but no tractors, and no tractors or work animals, United States, April 1, 1940* [1]

Value of product group	Percentage of farms having tractors	Percentage of farms having work animals but no tractors	Percentage of farms having no tractors or work animals	Total
Dollars	*Percent*	*Percent*	*Percent*	*Percent*
Under 250	4.7	42.6	52.7	100.0
250 to 599	7.7	64.4	27.9	100.0
600 to 999	16.4	68.4	15.2	100.0
1,000 to 1,999	36.7	55.3	8.0	100.0
2,000 to 3,999	61.3	33.7	5.0	100.0
4,000 to 9,999	75.7	19.0	5.3	100.0
10,000 and over	72.8	19.1	8.1	100.0
All groups	23.4	53.5	23.1	100.0

[1] See footnote 1, table 26.

About 1,054,000 farm operators, or nearly 18 percent of all operators, reported a total value of farm production ranging from $600 to $999. This group produced about the same total value of products as the lower two producing groups shown in table 26 combined. About two-thirds of the farmers in this group reported animal power only, and the other one-third was divided about equally between farmers having tractors and those owning no tractor or animal power. About one-fourth of these farms were subsistence farms in 1939.

The farms in these three low-producing groups made up nearly 4 million of the 6 million farms and produced only slightly more than one-fifth of the total value of farm production in 1939. These 3.9 million farms as a group were not highly mechanized in 1939 because very many of them were not physically and economically suited for mechanization. Many have very little cropland, many have rough topography, and many have unproductive soils. Operators of these farms are likely to get most of their income from livestock and livestock production from pasture lands and purchased concentrate feeds, and do not farm enough cropland to make their farms well suited to economical mechanization.

The fact that a few of these small farmers actually possess tractors (less than 10 percent) indicates that there may be others in this group who could afford to own tractors. Many other small farmers undoubtedly have some use of tractor power through exchange or custom work. The rapidity with which these small farms become mechanized through the buying of tractors and tractor equipment depends on several things. The relationship between farm prices and the prices of tractors and tractor equipment is important. The possibility of obtaining tractors and power machines that are physically and economically suitable for farms that have small acreages and low crop-producing capacity must be considered. Expansion of partnership operation of farms so that one set of power equipment can be used in place of two sets, or three sets, of smaller, horse-drawn machines may be feasible in some areas.

The availability of used tractors and tractor equipment may offer some small farmers an opportunity for greater mechanization of their farms, where cost of new machines is prohibitive. Ownership of tractors by small farmers for off-farm work, such as threshing, combining, and sawmilling, may expand in some areas. Garden tractors, it is expected, will be found on many more small farms. Home-made tractors may appear on more small farms as automobiles and trucks become available for conversion.

The effects of these four factors to date on the number of tractors on small farms cannot be measured, but it is believed that they have been considerable. On January 1, 1947 there probably were well over 100,000 garden-type tractors, and about 50,000 home-made tractors on farms of all sizes. Most of these probably were on the smaller farms. Manufacture of garden-type tractors for all purposes reached about 100,000 units in 1946. In previous recent years the average manufacture was around 10 to 25 thousand a year.

From the standpoint of promoting efficiency of production for an important segment of production the greatest opportunity for further expansion of tractors and tractor equipment is on the farms that had a production in 1939 valued at $1,000 or more per farm. In 1939 ther were about 2.1 million such farms. This one-third of our farms produced almost 80 percent of the total value of farm products; one-half of them reported having tractors in 1940. Many probably had more than one tractor each, but almost 7 percent of them reported no tractors or work animals.

The change from animal power to tractor power has been greatest in the northern agricultural regions. In 1939, for example, a very large part of total agricultural production in the northern geographic divisions came from farms reporting tractors, whereas a large part of production in the Southern States came from farms reporting animal power only (fig. 9). The East North Central and West North Central States combined produced 42.5 percent of the total United States value of products. Twenty-eight percent of this came from farms that had only horses and mules; and 72 percent from farms that had tractors or no animal or tractor power. A relatively large part of the production of the Middle Atlantic, Mountain, and Pacific Coast States came from farms having one or more tractors. Relatively large parts of the production in the New England, East South Central, and Pacific Coast States came from farms reporting no tractors or work animals.

These comparisons show the production from farms that reported whether they had or did not have farm tractors and work stock. This does not necessarily mean that farms reporting work animals only, or those reporting no work stock or tractors, do not have the use of custom or exchange tractors or of animals. The fact that about 1.4 million farms, or 23 percent of all farms in 1939, reported no tractor or animal power means, generally, that these farms are organized to operate without the actual ownership of these power units, rather than that their operations are underpowered. Operators of greenhouses and some commercial poultry enterprises who cultivate no land may have no reason to own tractors or work stock. Many fruit farmers in some areas hire all or a part of their field work done. On many such farms, motortrucks may represent the important power unit. Sharecropper farms in the South are operated with power and equipment owned by the "home farm."

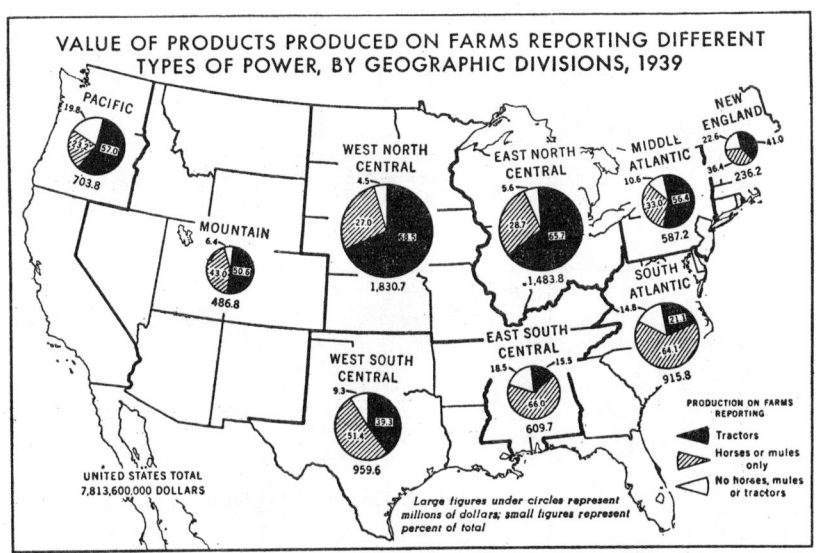

FIGURE 9—A LARGE PROPORTION OF THE VALUE OF TOTAL FARM PRODUCTS OF 1939 WAS PRODUCED ON FARMS HAVING TRACTORS IN THE MIDDLE ATLANTIC, NORTH CENTRAL, AND PACIFIC COAST STATES. RELATIVELY LITTLE OF THE 1939 PRODUCTION OF THE SOUTHEAST CAME FROM FARMS HAVING TRACTORS.

These are the chief reasons why 7 percent of the farms of medium size and 6 percent of the large farms reported no tractor or animal power. Some farms with products valued at considerably more than $10,000 reported no tractors, horses, or mules. Relatively large proportions of these medium and large farms were in the Pacific Coast States (table 28). A relatively large proportion of the small farms reporting no ownership of tractors or work stock were in the South Atlantic and East South Central States.

In 1940, 4.5 percent of all farm operators in the United States reported the ownership of tractors but no work stock. These farms were operated entirely, or almost so, with tractor power alone for field work. Some such farms were reported in each size group in each geographic division, and were of greatest relative importance on the larger farms. Relatively large numbers of tractor-operated farms were reported in the North Central States, the West South Central States, and in the

TABLE 28.—*Size of farm and percentage of all farms reporting different combinations of tractor and animal power, by value of product groups and geographic divisions, 1939* [1]

Geographic division	Percentage of all farms in U. S. in value group	Acres harvested per farm	Value of products per farm	Percentage of all farms in value group—			
				With horses and/or mules but no tractors	With horses and/or mules and tractors	With tractors only	With no tractors, horses or mules
	Percent	*Acres*	*Dollars*	*Percent*	*Percent*	*Percent*	*Percent*

FARMS WITH PRODUCTS VALUED AT LESS THAN $1,000

New England	2.0	13.6	362	0.5	0.1	0.1	1.3
Middle Atlantic	4.6	20.3	408	2.0	.4	.4	1.8
East North Central	12.8	24.1	442	6.1	1.5	.8	4.4
West North Central	13.0	53.9	482	6.6	2.2	.8	3.4
South Atlantic	19.5	18.9	439	12.7	.4	.1	6.3
East South Central	22.8	19.8	392	15.0	.3	.1	7.4
West South Central	18.7	29.1	439	13.3	.6	.5	4.3
Mountain	3.0	32.4	404	1.8	.3	.1	0.8
Pacific	3.6	10.7	393	1.0	.2	.3	2.1
United States	100.0	26.3	429	59.0	6.0	3.2	31.8

FARMS WITH PRODUCTS VALUED AT $1,000 TO $3,999

New England	2.3	35.6	2,066	1.2	0.4	0.2	0.5
Middle Atlantic	7.5	53.6	2,017	3.5	3.0	.4	.6
East North Central	24.0	75.9	1,939	8.9	13.1	1.2	.8
West North Central	27.1	137.2	1,975	9.4	15.4	1.4	.9
South Atlantic	12.5	46.7	1,691	9.9	1.2	.2	1.2
East South Central	6.3	54.7	1,620	4.6	.9	.1	.7
West South Central	10.9	106.1	1,736	5.6	3.0	1.6	.7
Mountain	4.5	102.8	2,016	2.5	1.4	.3	.3
Pacific	4.9	42.5	2,049	1.8	1.0	.8	1.3
United States	100.0	87.8	1,894	47.4	39.4	6.2	7.0

PROGRESS OF FARM MECHANIZATION

FARMS WITH PRODUCTS VALUED AT $4,000 OR MORE

New England	4.3	62.5	9,042	1.3	1.8	0.5	0.7
Middle Atlantic	9.5	88.7	8,317	1.7	5.6	1.4	.8
East North Central	19.5	176.5	7,263	1.4	16.4	1.3	.4
West North Central	26.6	247.3	7,713	3.0	21.3	1.7	.6
South Atlantic	7.1	159.1	9,440	2.5	3.4	0.5	.7
East South Central	3.2	179.3	8,266	1.2	1.8	(2)	.2
West South Central	9.3	272.1	10,626	2.5	5.2	1.2	.4
Mountain	8.3	239.9	10,653	3.0	4.2	0.7	.4
Pacific	12.2	201.8	12,336	2.3	5.0	3.2	1.7
United States	100.0	198.2	8,959	18.9	64.7	10.5	5.9

UNITED STATES TOTAL FOR ALL VALUE GROUPS

All geographic divisions	100.0	53.5	1,309	53.5	18.9	4.5	23.1

[1] See footnote 1, table 26.
[2] Less than one-tenth of 1 percent.

Pacific division. Some farms with both tractor and animal power were reported in each of the three size groups in each geographic division. Both tractors and work stock were used most extensively on the medium and large farms of the North Central States. Farms with horses and mules but no tractors were numerous in both the small and medium size groups, especially in the North Central and Southern divisions.

Any up-to-date summary of farmer use of tractor and animal power must consider the changes that have occurred since 1940. Data are not available for this purpose for different value-of-product groups, but are available for all farms as a group. From January 1, 1940 to January 1, 1945 the number of farms having tractors increased by approximately 607,000 and the number of tractors on farms increased by about 880,000. In this 5-year war period the increase in the number of farms having tractors was only 70 percent as great as the increase in the number of tractors on farms. This means that many farmers who already had one or more tractors took on one or two in addition. In 1940 the average for all farms reporting tractors was 1.1 tractors per farm, and in 1945, 1.2 tractors per farm.

This large wartime increase in tractor numbers raised the percentage of all farms that had one or more tractors from about 23 percent in 1940 to 33 percent in 1945. The wartime policy of increasing mechanical power and power machines on the larger and more productive farms probably means that 70 to 75 percent of the 2.1 million farms with production valued at $1,000 or more in 1939 had one or more tractors in 1945. As 7 percent of these farms reported no tractors or animal power in 1939, and supposedly are not generally in need of ownership of such power, only 20 percent, or about 400 thousand, did not have at least one tractor in 1945. Of the smaller farms, or those with a production valued at less than $1,000 in 1939, it is estimated that about 1.7 million had tractors, or no tractor or animal power in 1945. This leaves about 2.2 million small farms that had horses or mules but no tractors in 1945.

It now appears that tractors and tractor equipment on farms will continue to increase rapidly, as horses and mules continue to disappear and as new and more adaptable machines are manufactured. If tractors continue to increase during the next 5 years as fast as between 1940 and 1945, about 2.6 million farms will have tractors in 1950. If each of these farms has an average of about 1.2 tractors (the average per farm in 1945), there will be approximately 3 million tractors on farms at that time.

Mechanization on farms best suited for tractors and tractor equipment has gone a long way. If farmers continue to buy and keep tractors as indicated above, it might be that one or more tractors will be found on 90 percent of the farms that reported products valued at $1,000 or more in 1939, and on 18 percent of the farms that reported products valued at less than $1,000.

On January 1, 1947 there were about 2.7 million tractors on farms. Many of these were above normal average age because a large part of the wartime increase in tractor numbers came from the reconditioning of tractors that ordinarily would have been discarded. Most of these reconditioned tractors were put into use after 1942, and it is believed that the majority of them are still on farms. It is estimated that as many as 400,000 of the tractors on farms on January 1, 1947 normally would have been discarded by that date. Under favorable conditions another 400,000 might be available for discarding between January 1, 1947 and January 1, 1950. If these discards are actually made, there will be need for 800,000 tractors, plus another 400,000 tractors needed to increase the total number from 2.7 million in 1947 to around 3 million in 1950, or a total of 1.2 million new tractors between January 1, 1947 and January 1, 1950. These possibilities are in terms of field tractors, and do not include garden tractors, although the number of these probably will increase rapidly.

Growing Importance of Farm Electrification

At the middle of 1946 more than half the 6,000,000 farms of the United States had central-station electric power service. This is more than double the number that were so electrified in 1939, and nearly 5.5 times the number that had such service in 1930 (table 29). The rural electrification movement of the last 10 years is advancing rapidly.

TABLE 29.—*Farms in the United States with central-station electric service, 1925–46* [1]

Year	Farms electrified, January 1		Index of farm electrification growth (1936=100)
	Number	Percentage of total farms	
1925	204,780	3.2	26.0
1930	576,168	9.0	73.0
1935	743,954	10.9	94.3
1936	788,795	11.6	100.0
1937	1,042,924	15.3	132.2
1938	1,241,505	18.2	157.4
1939	1,406,579	20.6	178.3
1940	1,786,000	26.2	226.4
1941	2,069,759	33.9	262.4
1942	2,352,603	38.6	298.3
1943	2,486,230	40.8	315.2
1944	2,557,247	41.9	324.0
June 30, 1945	2,725,610	44.7	345.5
June 30, 1946	3,106,775	52.9	393.9

[1] Data for 1925–44 from Rural Electrification After the War, United States Department of Agriculture, A.I.S.–11, Feb. 1945; data for June 30, 1945 and 1946 from reports of Rural Electrification Administration, United States Department of Agriculture.

In 1943 the estimated value of major electrical equipment on farms in the United States was nearly 625 million dollars. About 63 percent of this value was in home appliances, including such items as radios, irons, washing machines, refrigerators, toasters, vacuum cleaners, and hot plates. Thirty-seven percent was in farm equipment, principally for the operation of brooders, water pumps and installations, miscellaneous electric motors, cream separators, milking machines, and milk coolers.

Thus, although far more electrical equipment is now found in rural homes and in farm barns, as yet it has had no appreciable effect on the principal farm machines and tools used in crop production. Widespread rural electrification is so new that a cumulative demand for electrical equipment on farms that are already electrified will continue for some time. More ways will be found for using it, especially in and around buildings. On July 1, 1945, Michigan, New Jersey, and California led all States, with nearly 90 percent of the farms receiving electric service. More than 80 percent of the farms in several other States in the Eastern and Western parts of the United States were receiving electric service. Relatively small proportions of the farms in the Great Plains and Southern States had electric service in 1945 (see table 46 in the appendix).

MECHANIZATION AND PRODUCTION COSTS AND RETURNS

Mechanization and other technological developments usually enable farmers to produce more product with an hour of labor and a dollar's worth of power and equipment. Other things being equal, these efficiencies would result in lower production costs per unit of product, and this would mean greater profits to farmers or lower food prices in the markets, or both. The facts are, however, that "other things" do not remain the same. Prices of the things that farmers buy and prices of the things they sell both change constantly. Rents and wages change in agriculture, as in industry. Transportation and handling margins of distributors and retailers are by no means fixed over a period of time. The result is that although physical costs of each unit of farm output have decreased about one-fourth during the last quarter of a century, current money costs and prices of farm products have fluctuated so violently that there have been no definite sustained trends in current costs or real labor return per unit of output.

Total costs of agricultural production as affected by mechanization are here treated historically for the period 1910-45. Physical production costs are in terms of quantities multiplied by 1935-39 average prices, and current costs are in terms of quantities multiplied by current calendar-year prices. Differences between the two are the result of differences in prices of cost factors. Some of the comparisons are affected by the procedures used in computing costs of agricultural production and by conditions during the base period. These basic considerations are discussed early in this analysis. Special attention is given to segregating important operating-cost items that bear directly on the effect of mechanization on production costs. As much of the residual income (after cash costs) accrues to the farm operator and family labor, particular attention is given to the "real return" of farm labor, and to the influence of the labor item on operating costs of tractors, trucks, automobiles, other machinery, and horses and mules. Comparisons are made of net returns to the operator and family labor, and of wages paid to hired labor. The interchangeability of machinery and labor, and of power machines and animal-operated machines, in relation to production cost efficiencies, are analyzed.

The Base Period

The 1935-39 base period was used in the analysis of farm mechanization and production costs and returns. This period seems better suited in most respects for the analysis than other base periods that are sometimes used in constructing series of index numbers. The years 1935-39 constitute the most recent prewar combination of 5 years, and a recent base is preferred to an old base. This period is used extensively as a base by the Bureau of Agricultural Economics and other agencies in constructing various series of index numbers which were useful in this study.

Two other base periods commonly used are 1910-14 and 1925-29. Table 30 compares pertinent economic conditions in the 1935-39 period with those in the other two commonly used periods. This comparison is made for the primary purpose of giving the reader an understanding of the relative economic position of farmers in 1935-39 and at two other important times during the 36-year period. This seems highly desirable to a proper understanding of the effects of mechanization on the welfare of farm people, as measured by costs and returns in agriculture.

TABLE 30.—*Relative economic position of farmers in the United States in specified base periods*

Item	Unit	Average of—		
		1935-39	1925-29	1910-14
Prices received by farmers	Index, August 1909–July 1914=100.	107	149	100
Prices paid by farmers including interest and taxes	Index, 1910-14=100	128	168	100
Ratio of prices received to prices paid	...do...	84	89	100
Farm employment	Millions	10.9	11.4	12.1
Farm wage rates	Index, 1910-14=100	118	179	100
Average value per acre of farm real estate	Index, 1912-14=100	83	121	[1] 100
Ratio of net land rent to land value	Percent	5.1	6.0	[1] 4.4
Farm-mortgage interest rate	...do...	5.0	6.1	6.1
Short-term interest rate	...do...	6.3	7.6	8.7
Crop yields per harvested acre[2]	Index, 1910-14=100	105	99	100
Farm output per worker	...do...	139	128	100
Labor returns per farm worker	Current dollars	386	464	258
Labor returns per operator and family worker	...do...	394	475	254
Labor returns per hired farm worker	...do...	362	433	271
Real labor returns per farm worker	Average 1935-39 dollars	386	354	318
Real annual wages per industrial worker[3]	...do...	956	1,033	815
Ratio: Real labor returns per farm worker to real annual wages per industrial worker[4]	Percent	40	34	39

[1] Average of 1912-14.
[2] Combined yields of 18 field crops and 10 fruit crops.
[3] Adjusted for unemployment.
[4] For an explanation of the apparent relatively low returns to farm labor in all 3 periods see text, p. 73.

Both prices paid and prices received by farmers were higher in 1935–39 and in 1925–29 than in 1910–14. Prices paid increased faster than prices received so that the ratio of prices received to prices paid was less favorable in 1925–29 than in 1910–14, and still less favorable in 1935–39. Counting 1910–14 as 100, the ratio in 1925–29 was 89, and the ratio in 1935–39 was only 84. The less favorable position of prices received, when compared with prices of cost inputs, in 1935–39 is much less significant than seems apparent from this comparison alone. The number of farm workers was lower and farm output per worker was higher in both 1925–29 and 1935–39 than in 1910–14. The result was that even though prices of farm products in the base period used here were relatively low compared with those of the other 2 base periods, real labor returns per farm worker were higher than in 1910–14 by 11 percent in 1925–29 and by 21 percent in 1935–39. This fact is one of the compelling reasons for using a recent base period in the sort of analysis made in this part of the publication.

Yields of crops per harvested acre were about the same in 1910–14 and 1925–29. Yields in 1935–39, however, averaged 5 percent higher than in the other two periods, despite the adverse effect of the 1936 drought. Good growing conditions in the last 3 years of the 1935–39 period more than offset the low yields in 1936. For the United States as a whole, the period 1935–39 on the average did not represent extreme growing conditions.

Other comparisons contained in table 30 show that land values per acre in the United States were about a fifth higher in 1925–29, and 17 percent lower in 1935–39 than in 1912–14. Farm-mortgage rates and short-term interest rates were also lower in the two later base periods than they were in the 1910–14 period. On the other hand, land rents were higher in relation to land values in both recent periods than in 1912–14, averaging 36 percent higher in 1925–29, and 16 percent higher in 1935–39. These increases are in the same direction as increases in real labor returns per worker shown above.

Farm operators and members of their families received on the average about 10 percent more a year for their work than hired workers in 1935–39 and 1925–29, but in 1910–14 hired workers were paid somewhat more than the farm operators and family workers had left for their work. The ratio of real labor returns of all farm workers to real annual wages of industrial workers was about the same in both 1935–39 and 1910–14, but the ratio was less favorable to farm workers in 1925–29.

All in all, the period 1935–39 can be characterized as one that represents a middle field of economic relationships in farming. Farmers and farm workers were not extremely bad off nor were they exceptionally well off. The 1935–39 period includes recent items of mechanization and production, and takes into account recent influences of physical efficiencies in production. From this standpoint it becomes more acceptable than earlier periods in any analysis of costs and returns in agriculture, as affected by mechanization.

Importance of Power and Machinery Costs

The production costs used herein are net costs to agriculture in the aggregate. (See footnote 1, table 31). The costs include a land charge.

a charge for family and operator labor, cash expenditures for hired labor, fertilizer and other items not produced by farmers, and allowances for depreciation of buildings and equipment. Labor of the farmer and his family was charged to agriculture on the basis of prevailing

TABLE 31.—*Total production costs of agriculture, United States, 1935–39 average* [1]

Item	Farm labor [2]	Other costs [3]	Total costs	
			Amount	Proportion of all costs
	Million dollars	Million dollars	Million dollars	Percent
Tractors	15	344	359	4
Trucks	10	209	219	3
Automobiles [4]	20	333	353	4
Other machinery	58	441	499	6
Farm-produced power	341	699	1,040	12
Other farm labor [5]	3,014		3,014	36
All labor, power and machinery	3,458	2,026	5,484	65
All other costs [6]		2,973	2,973	35
Total	3,458	4,999	[7] 8,457	100

[1] The total costs shown in this table cover about 97 percent of all production costs, cash and noncash. Noncash items were valued at cash-cost rates. A rental charge for all farm land was estimated from total net rent paid on land rented and the proportion that the value of rented land was of the value of all farm land. Several types of data were utilized in estimating the value of unpaid operator and family labor—the hired-wage bill, estimates of man-hour requirements of farm production, hired and family farm-employment and farm-wage rates—in order to charge the same rate per man-equivalent hour for both hired and unpaid labor. Interest charges on property other than real estate were calculated from value of investment and current interest rates on short-term credit. Annual depreciation charges for machinery and buildings were used instead of annual cash expenditures.

The costs shown are substantially "net costs to agriculture". For example, costs of feed and livestock bought by farmers were not included, since such items are largely sales by one farmer to another. Double counting would occur if such costs had been included since the cost of the feed to agriculture is already reflected in the labor, machinery, rent, fertilizer, and other items included as costs.

Most of the data upon which the cost estimates were based are contained in the BAE processed report, *Net Farm Income and Parity Report: 1943* and summary for 1910–42, July 1944.

[2] Estimated costs of farm labor used in the repair, maintenance, and housing of motor vehicles and machinery were allocated to these items. Labor costs for chores, housing, and raising of horse and mule feed were estimated for farm-produced power.

[3] Other costs of power and machinery items include depreciation, operating expenses, repairs, skilled nonfarm labor, insurance, and allowances for taxes, interest, and housing. Machinery and power costs of raising horse and mule feed are included in the respective power and machinery items but are excluded from costs of farm-produced power.

[4] Only the farm-production share of automobile costs is included.

[5] Includes costs of all farm labor other than those allocated to power and machinery costs.

[6] Includes rent, interest, fertilizer, and other cost items not allocated to power and machinery costs.

[7] The total costs of 8.5 billion dollars shown here are 1 billion less than the adjusted gross income to agriculture for this period. (See footnote 9, page 69 for an explanation of adjusted gross income). About one-third of this difference of 1 billion dollars results from an incomplete measurement of total costs. The remaining difference of two-thirds billion dollars means that operators and family workers received 5 cents more labor returns per hour of work than hired workers received during the 1935–39 period. Wage rates paid to hired workers were used in calculating total labor costs.

rates for hired labor. The net costs to agriculture represent what all farmers as a group would have to recover in cash expenses paid to those not engaged in farming, and to cover depreciation of buildings and equipment, going wages for operator and family labor, customary short-time mortgage rates for working capital, and average rentals paid by tenants for the use of land. The costs, therefore, include not only cash expenses, but allowances for non-cash production items as well.

Computed in this way the total production costs of agriculture averaged during the base period, 1935–39, about 8.5 billion dollars (table 31). Nearly 3.5 billion dollars of the total was for farm labor, approximately 27 percent of which was hired; about 5 billion dollars were for all other production costs. About 13 percent of the cost of all farm labor in the base period was for farm-produced power and maintenance of machinery, and 87 percent represented all other farm labor costs. Of the power and machinery labor costs, nearly 80 percent was for farm-produced power, including labor costs for producing horse and mule feed and caring for the work stock and colts; only 20 percent was for farm servicing of tractors, trucks, automobiles and other machinery and tools. All labor, power, and machinery costs in agricultural production in the base period amounted to nearly two-thirds of all production costs.

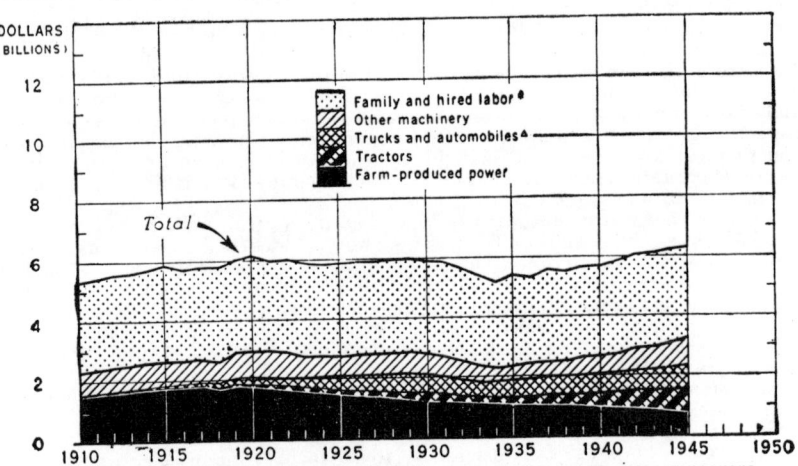

FIGURE 10—TOTAL PRODUCTION COSTS OF FARM POWER, MACHINERY, AND LABOR, UNITED STATES, 1910-45.
(COSTS IN 1935-39 AVERAGE DOLLARS)

In terms of 1935–39 average dollars, total operating costs of farm power, machinery, and labor have varied from a low of 5.2 billion dollars in 1934 to a high of 6.3 billion dollars in 1945, a range in 36 years of only about 20 percent. More costs of operating power machines have made up for reductions in costs of work stock.

Measured in terms of 1935–39 average dollars, the total physical costs of farm labor, power, and machinery have remained remarkably stable over the 36 years, 1910–45, despite a large increase in farm production. The interwar peak of these physical costs was reached in 1920 when inventories of horse-drawn and power machines were piling up, and the low point was reached in the depression year, 1934, when machinery purchases and operating costs were being reduced drastically (fig. 10). The peak volume of physical costs reached in 1920 was exceeded slightly in 1944 and 1945.

Millions of horses and mules have been replaced since World War I by motor vehicles and modern power-operated machines without any appreciable increase in total operating costs of power and machinery. But the change in composition of farm power and machinery has made possible a large increase in farm production with fewer man-hours of labor.

During 1910–14 about one-fifth of all man-hours used on farms were devoted to producing horse and mule feed and to caring for horses and mules and servicing farm machinery and equipment. This meant that more than one day out of every week of work, exclusive of Sundays, was spent on such work. During 1942–45, total man-hours of farm labor used on all farm power and machinery made up only slightly more than 11 percent of the total hours in agriculture, an equivalent of two-thirds of a day out of each week of 6 work days (table 32). This decrease is due largely to the decline of farm-produced power. Horses and mules in 1945 required only about 1.4 billion man-hours, including hours for producing their feed, compared with an average of 3.7 billion hours in 1910–14.

Effect of Prices on Production Costs

The rather stable production costs of farm power, machinery, and labor shown in figure 10 in terms of 1935–39 prices become decidedly unstable when expressed in current dollars as in figure 11. Sharp ups and downs are evident in production costs because of high and low prices of power and machinery cost factors in periods of war and depression. The range in current dollar costs is from a low in 1933 of less than 4 billion dollars to a high in 1945 of 13.4 billion dollars, a difference of 250 percent compared with a range from low to high of only 22 percent when computed at average 1935–39 prices.

Fluctuations were much less violent in the current operating costs of machinery than in the operating costs for farm-produced power and farm labor. High farm wages and high prices of horse and mule feed in the war periods, and low wages and feed prices in the depression periods are largely responsible for the tremendous fluctuation in total current costs for all farm labor, power, and machinery.

The labor cost shown in these calculations includes the return that farm operators and their families would have received if they had been paid for their farm work at prevailing wage rates for hired farm workers. When wage rates paid hired workers were low *all* workers on farms were receiving low returns for their efforts, and when wage rates were high *all* workers on farms were getting higher returns for their contribution to production. A part of the labor force was actually hired at these rates by farm operators, but usually about 75 percent of the work was done by operators and members of their families. They received their labor returns from net farm income after other expenses were paid.

TABLE 32.—*Farm labor requirements for care and maintenance of horses and mules, motor vehicles, and machinery, and labor requirements for other farm work, United States, specified periods and years, 1910–45*

Period or year	Farm-produced power			Servicing and maintenance of—		All power and machinery	All other farm work	Total farm work	Proportion of total farm work required for all power and machinery
	Growing horse and mule feed	Chore and overhead work	Total	Tractors, trucks, and automobiles [2]	Other farm machinery				
	Million hours	*Million hours*	*Million hours*	*Million hours*	*Million hours*	*Million hours*	*Million hours*	*Million hours*	*Percent*
Average of:									
1910–14	1,714	1,956	3,670	7	589	4,266	18,240	22,506	19
1915–19	1,711	2,092	3,803	42	569	4,414	18,581	22,995	19
1920–24	1,598	1,952	3,550	117	599	4,266	18,507	22,773	19
1925–29	1,311	1,706	3,017	199	485	3,701	18,657	22,358	17
1930–34	1,085	1,392	2,477	236	440	3,153	18,234	21,387	15
1935–39	919	1,143	2,062	262	345	2,669	17,904	20,573	13
1940–44	646	961	1,607	329	478	2,414	18,460	20,874	12
1940	736	1,021	1,757	295	409	2,461	17,951	20,412	12
1941	699	996	1,695	309	430	2,434	18,183	20,617	12
1942	623	967	1,590	331	474	2,395	18,737	21,132	11
1943	591	932	1,523	349	527	2,399	18,627	21,026	11
1944	582	889	1,471	361	551	2,383	18,799	21,182	11
1945	550	843	1,393	381	613	2,387	18,268	20,655	12

[1] See footnote 2, table 5.
[2] Includes farm labor on only that share of automobiles used in farm production.

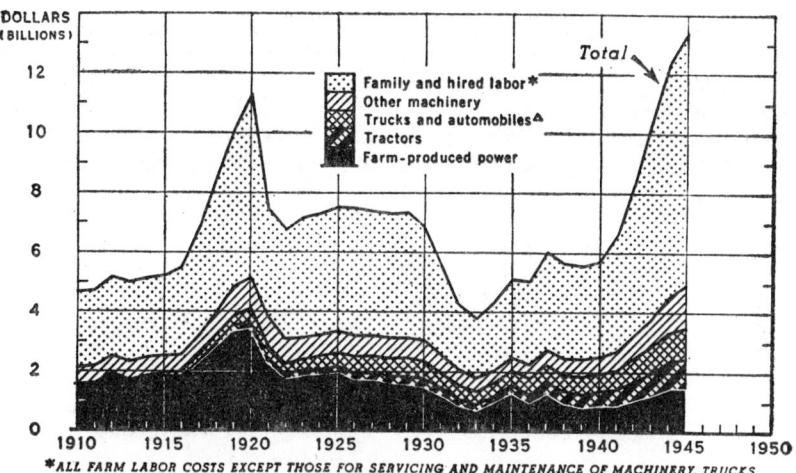

FIGURE 11—TOTAL PRODUCTION COSTS OF FARM POWER, MACHINERY, AND LABOR, UNITED STATES, 1910-45.
(COSTS IN CURRENT DOLLARS)

Wide fluctuations in operating costs of power, machinery, and farm labor are caused by severe changes in prices of these cost factors rather than by changes in volume of the cost factors. (See figure 10).

The first 3 sections in figure 12 emphasize the shift in composition of power costs. Aggregate costs of farm-produced power in terms of 1935-39 average prices have declined rapidly since 1919, but as the volume of farm output has increased over the period, costs per unit of output have declined faster than total costs. On the other hand, total physical operating costs of tractors, and trucks and automobiles, have gone up since 1910, with physical costs per unit of farm output increasing less rapidly than total costs.

The fourth section in figure 12 is for machinery other than tractors, trucks and automobiles. Even though the composition of this machinery item has changed drastically, the index of operating cost per unit of output was decidedly downward until just before the recent war.

In section 5 of the illustration the indexes of aggregate operating costs of farm-produced power and other machinery have been combined with the indexes of operating costs of tractors, automobiles, and trucks. The result is a relatively flat trend in total physical costs since World War I but a decided downward trend in the index of production costs per unit of farm output. This comparison emphasizes the increase in physical efficiency of farm power and machinery because of the marked increase in volume of farm output and the change in composition of machinery and power.

Section 6 of figure 12 brings out the effect of substitution of machinery and power for labor. Thus, when physical costs of all farm labor are combined with those for power and machinery, the total costs show virtually a flat trend over the entire period. The strong upward trend in farm output, with practically a constant volume of production costs for labor, power and machinery, has brought a reduction of physical costs per unit of output of about 33 percent since World War I (1919–1945). This means that farmers as a group in 1945 were using one-third less total labor, power, and machinery combined, to produce one unit of farm output than was used in 1919. A part of this reduction in physical costs per unit must be ascribed to nonfarm workers who have contributed indirectly to these reductions. The sharp fluctuations in costs per unit of production in 1934 and 1936 were caused by low production during the unusually severe droughts. The notable fluctuations in the World War I period were caused primarily by yearly variations in volume of farm output and in volume of farm-produced power and machinery available to farmers.

The production costs in 1935–39 dollars (fig. 12) are reproduced in current dollars in figure 13. Comparison of the corresponding sections

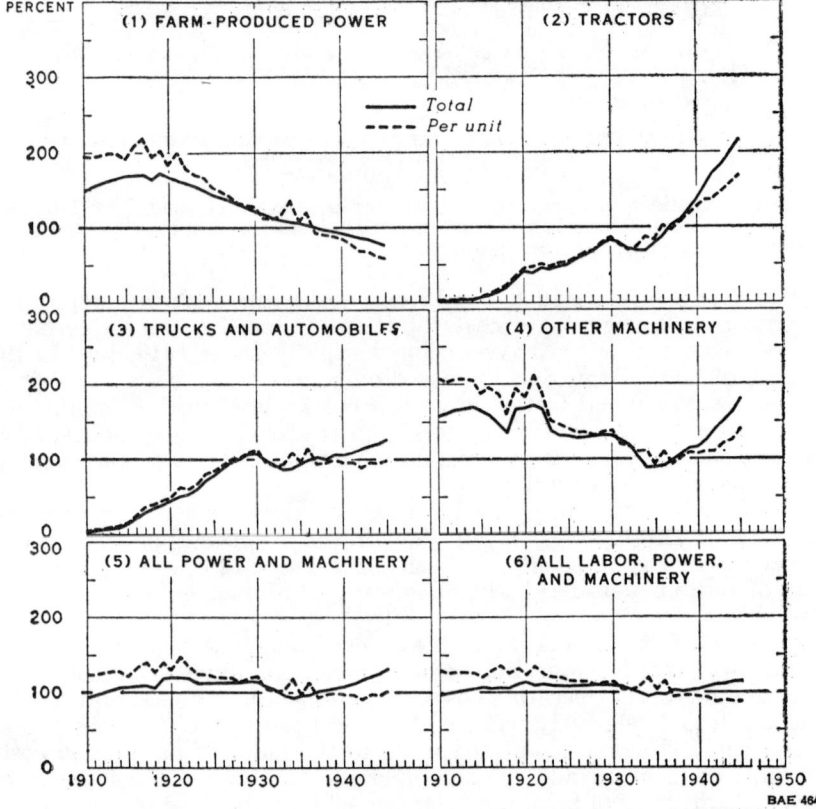

FIGURE 12—TOTAL PRODUCTION COSTS AND COSTS PER UNIT OF FARM OUTPUT FOR FARM POWER, MACHINERY, AND LABOR, UNITED STATES, 1910–45. (COSTS IN 1935-39 AVERAGE DOLLARS; INDEX NUMBERS 1935-39=100)

Total physical costs of farm-produced power have decreased while physical costs of tractors and automobiles and trucks have increased. When these are combined with total costs of other machinery into a single index the result is a relatively flat trend in the total but a decided downward trend in the cost index per unit of farm output. The addition of physical costs of farm labor results in a flatter trend for the total and a sharper downward trend in physical costs per unit of farm output.

of the two emphasizes the strong influence of prices and of farm-produced power on production costs of labor, power, and machinery. Section 2 of figure 13 shows that the long-time upward trend of current dollar operating costs of tractors does not differ greatly from the trend of costs in constant dollars. The relatively smooth upward trend in costs of operating tractors, therefore, was largely the result of increased volume, although prices of cost factors had a noticeable effect on total costs in the two war periods and during the depression of the 1930's.

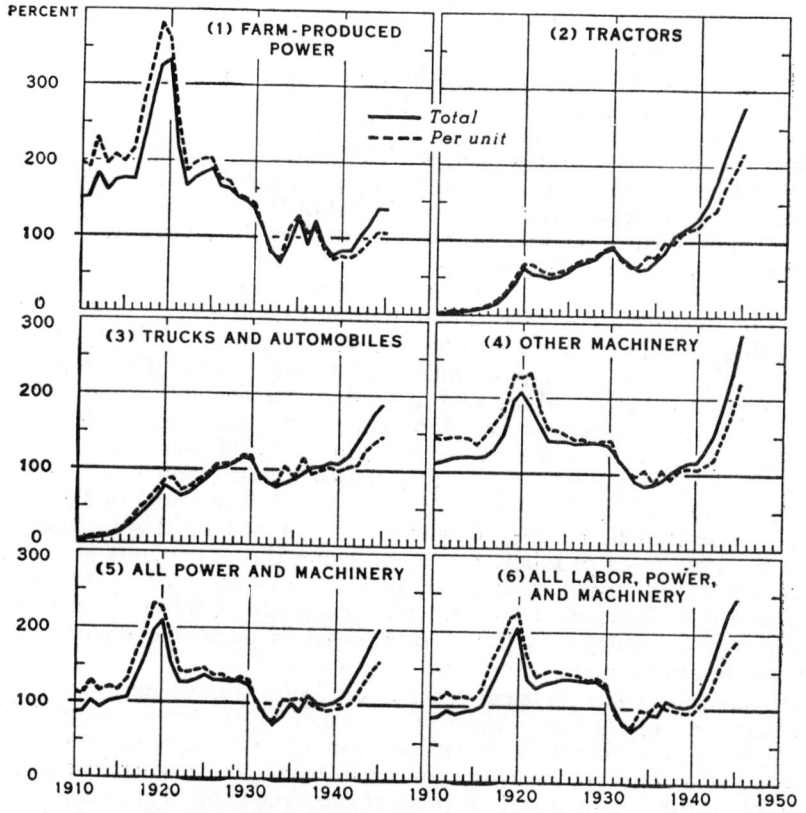

FIGURE 13—TOTAL PRODUCTION COSTS AND COSTS PER UNIT OF FARM OUTPUT FOR FARM POWER, MACHINERY, AND LABOR, UNITED STATES, 1910-45. (COSTS IN CURRENT DOLLARS; INDEX NUMBERS 1935-39=100)

These current dollar costs when compared with constant dollar costs (fig. 12) bring out the strong influence of prices on production costs of labor, power, and machinery. Operating costs of tractors, and automobiles and trucks, have been influenced less by price changes than have costs of farm-produced power, other machinery, and farm labor.

In contrast to operating costs of tractors, costs of farm-produced power have been influenced rather severely by price changes. Consequently, when plotted in current-dollar costs they do not show the smooth trend shown when plotted in constant-dollar costs. The indexes of total current-dollar costs of farm-produced power varied more than 400 percent from a high in 1920 to a low in 1933. The higher costs in 1920 were due to more animals and to higher maintenance costs. Wide variations in feed prices and wages of farm labor were principally responsible for more than one-half of this wide variation in total costs of farm-produced power. In terms of constant dollars, the variation for the same period was about 150 percent.

The upward trend in the combined operating costs of automobiles and trucks, based on current prices, was largely the result of increasing numbers up to 1930, although prices influenced total operating costs of these machines considerably in the two war periods. Changes in costs of operating "other machinery" in the period between the wars were mainly brought about by changes in volume of machinery.

The strong influence of costs of farm-produced power on the costs of all power and machinery is shown by the similarity of data represented by sections 1 and 5 in figure 13. When current labor costs are added to current costs of all power and machinery, the general picture is not changed much, except for the period of World War II when relatively high farm wages boosted the total costs substantially. The striking dissimilarity between operating costs of all labor, power, and machinery in current and constant dollars emphasizes the influence of prices on operating costs (section 6, figures 12 and 13).

During the base period, 1935–39, labor, power, and machinery costs made up 65 percent of all costs to agriculture. The dominant influence of costs of labor, power, and machinery on total costs is evident in periods of war and depression, as in the base period. This influence can be seen by noting the similarity of the indexes of all costs, and those for labor, power, and machinery (shown in section 6 of figure 12 and section 1 of figure 14; and in section 6 of figure 13 and section 2 of figure 14). This similarity is striking despite the fact that physical costs of labor, power, and machinery per unit of output have decreased faster than the physical costs of some other items. For example, from 1919 to 1945 physical costs of labor, power, and machinery per unit of output were reduced 33 percent compared with a reduction of 26 percent in all physical costs.

Historically, total physical costs in agriculture have been decidedly constant during the last quarter of a century. The decrease in physical unit costs of one-fourth has resulted from an increase of more than 40 percent in volume of farm output with an accompanying increase in total operating costs of only 8 percent (section 1, fig. 14). This means that if there had been no increase in farm-production efficiency during the last 25 years, total cost for the larger volume of production in 1945 would have been 33 percent higher than it actually was. It means, also, that most of the increased efficiency in agricultural production when expressed in terms of production per unit of input has been the result of greater total production rather than a result of any decrease in total volume of production inputs.

Over the third of a century shown, prices of cost goods have fluctuated so violently that price changes have over-shadowed the changes in physical efficiency in agriculture as a whole. Annual average farm wages per day without board, for example, varied during the 36 years from $1.11 in 1933 to $4.34 in 1945. Farm prices of corn, an influential item in the cost of farm-produced power, varied from about 32 cents per bushel in 1932 to about $1.52 in 1918. But despite price changes farmers who reduced their costs have had larger net incomes than they would have had without reductions in costs.

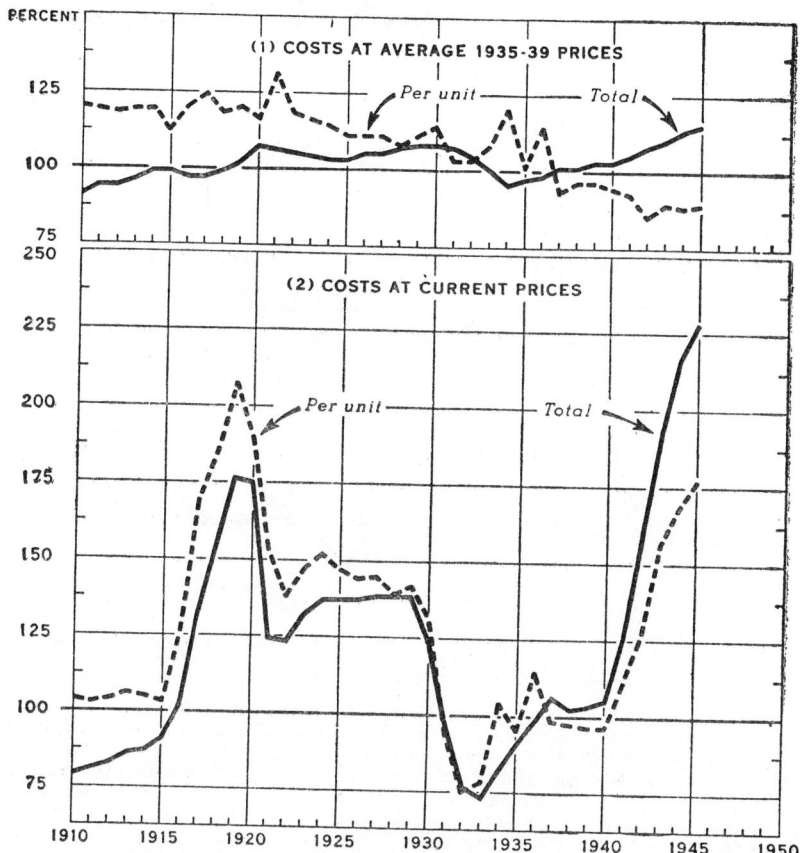

FIGURE 14—TOTAL PRODUCTION COSTS AND COSTS PER UNIT OF FARM OUTPUT, UNITED STATES, 1910-45.
(INDEX NUMBERS, 1935-39=100)

As about two-thirds (65 percent in 1935-39) of all operating costs in agriculture are for labor, power, and machinery, these items are dominating factors in the trend of all costs. This influence may be observed by comparing the indexes of all costs shown here with the indexes of labor, power, and machinery costs shown in figure 13.

An understanding of the dominating influence of prices of cost factors on unit production costs can be had by comparing sections 1 and 2 of figure 14. For example, section 2 shows that the difference in costs per unit of farm output between the high point in 1919 and the low point in 1932 was 132 index points, but when the influence of price changes was eliminated, as in Section 1, the difference was only 17 points. From 1932 to 1945 the trend in current production costs was reversed, and total cost per unit of output increased rapidly, chiefly because of increasing prices rather than because of the use of increasing

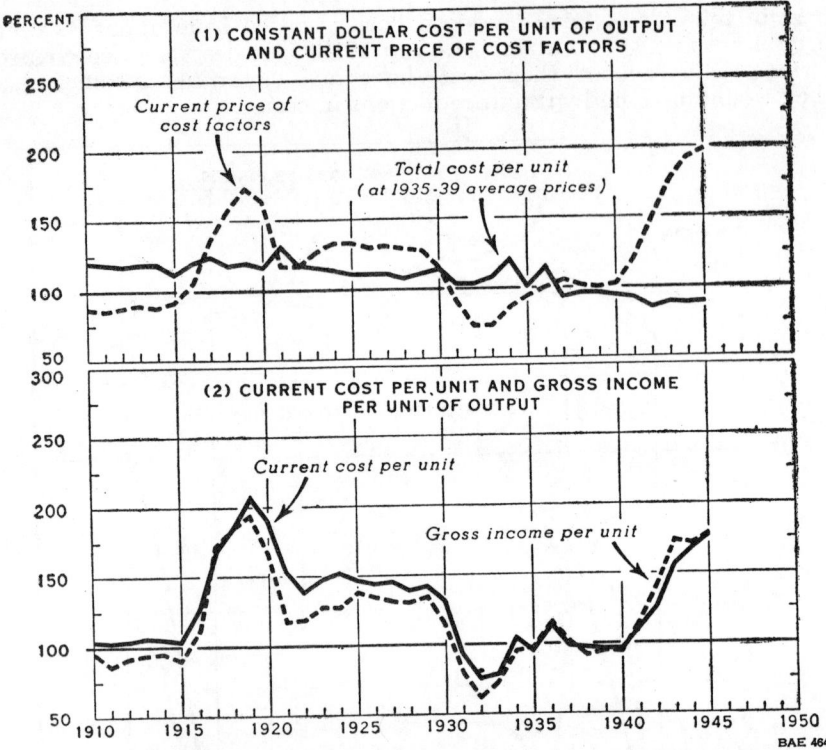

FIGURE 15—RELATION OF PRICE OF COST FACTORS AND PRODUCTION COSTS TO GROSS INCOME PER UNIT OF OUTPUT, UNITED STATES, 1910-45.
(INDEX NUMBERS 1935-39=100)

Fluctuations in current prices of cost factors are responsible for the wide variations in current costs per unit of output. In the World War II period, current-dollar costs per unit of output increased less than prices of cost factors because of increased production per worker and per unit of machinery. The close relationship between land rent and farm wages, and farm prices of farm products, is largely responsible for the close relationship between current costs and gross income per unit of output. When costs per unit are higher than gross income per unit, many farm operators receive less for their labor and/or use of land than the wages paid to hired labor and the rents paid by tenants.

quantities of cost goods per unit of output. The increase in current dollar costs amounted to 102 index points compared with a decrease in 1935–39 average dollar costs of 14 index points. The increase in prices of all cost factors combined [8] from the depression year of 1932 to the war year of 1945 was 126 index points, or about 170 percent (section 1, fig. 15). So long as prices of input factors are influenced so strongly by periods of prosperity and depression we cannot expect that a decreasing trend in physical costs will be paralleled by a decreasing trend in dollar costs.

---

[8] The index of current prices of cost factors was calculated by dividing the index of total current costs by the index of total costs at 1935–39 average prices. As the index of costs at 1935–39 average prices is essentially a weighted-average measure of volume of physical inputs, this method gives a measure of weighted-average price changes of labor, power, machinery, fertilizer, rent, etc.

Other aspects of the farm-cost problem should be mentioned here. In the first place, the net returns that farmers get for their products depend upon prices of the products as well as upon prices of cost goods. This phase of the problem is discussed under the next heading, Agricultural Costs and Returns. In the second place, as production costs include wages for hired labor and for farm operators and family workers, changes in wage rates reflect changes in labor returns to farm people as well as changes in production costs. Individual operators who have reduced their cash operating costs, or increased their output per unit of cash cost, have obtained higher net incomes, or have had more left as wages for themselves and their family workers, than they would have had with no increase in operating efficiency.

Increased physical efficiency in agricultural production has been a fairly persistent and stable force over a long period, tending always to decrease the cost of production per unit of output. On the other hand, prices of the things farmers buy or furnish for production purposes have fluctuated so violently that increased production efficiency has not at all times meant a corresponding increase in income, or in returns for the labor of farm people.

Agricultural Costs and Returns

Prices received by farmers for the products they sell are also subject to wide fluctuations, as are the prices of things they buy. Price levels of both, therefore, are determined largely by over-all economic conditions, the greatest fluctuations occurring in periods of war and depression. Although prices of the things the farmer buys may respond more slowly to economic conditions than prices of the things he sells, in general, changes in prices received and changes in production costs per unit move together. This close relationship is brought out in section 2 of figure 15, where current costs per unit of output are compared with average prices per unit, as measured by gross income per unit.[9]

One outstanding reason for the general tendency for total costs to be low in periods of low income and high in periods of high income is that income received greatly influences the farm wages and land rents that will prevail. These cost items are largely "imputed costs," and are at the same time largely "imputed returns" to farm people for their labor and their land.[10] During the base period these two items made up more than 60 percent of the imputed total costs to agriculture as computed here.

Because a large part of the costs of rent and labor to agriculture represents what is left for the farm families' labor and investment

---

[9] The index of gross income per unit of output was computed by dividing the index of gross income by the index of farm output. The index of gross income was constructed from the gross income figures reported by BAE, adjusted to make them comparable in content to the production cost figures. BAE gross income includes cash income from marketings of crops and livestock, plus the value of farm products consumed in the home, plus the rental value of farm homes, plus amounts received in Government payments. This total was adjusted by subtracting costs of feed and livestock bought by farmers, and by adding or subtracting plus or minus changes in inventories.

[10] Not all of the land is owned by farm people; nor is all of the work done by farm workers but, broadly speaking, most of the land and labor costs constitute returns to farm people.

after paying cash operating expenses and charging off depreciation, "low costs" that are brought about by low wages and rents frequently mean low income to agriculture. Low costs of this kind work in the opposite direction from low costs that are the result of increased efficiency in production. Furthermore, the volume of costs of labor and rent is so large, and the fluctuations in these costs are so great that these imputed costs dominate the cost and returns situation in agriculture.

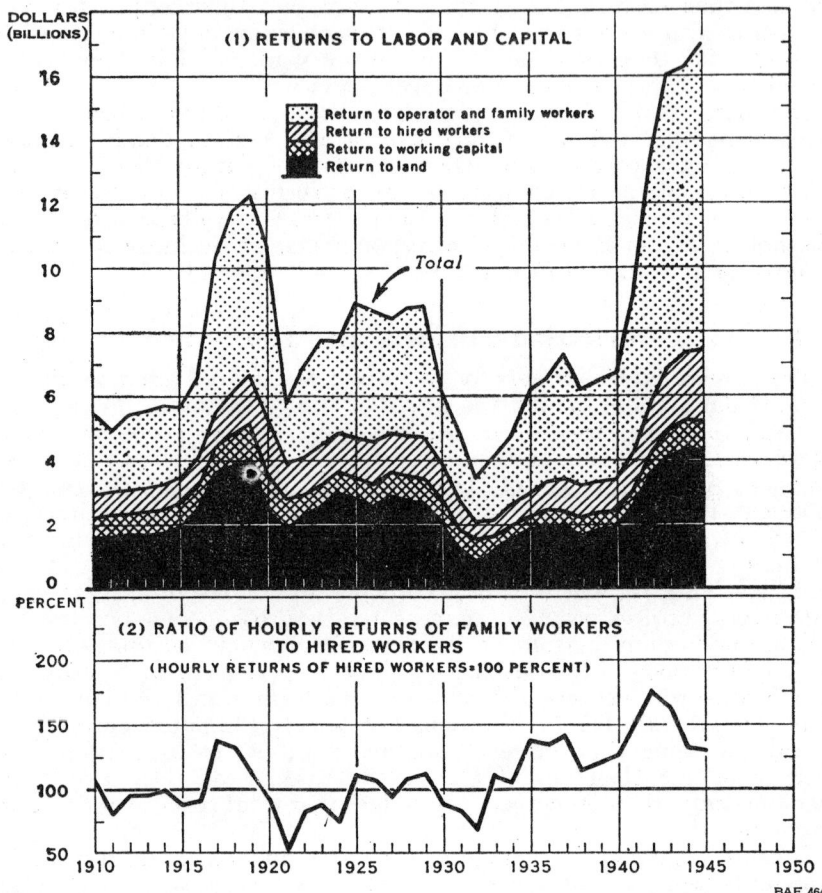

FIGURE 16—RETURNS TO ALL FARM LABOR AND CAPITAL, AND RELATIVE RETURNS TO FAMILY AND HIRED LABOR, UNITED STATES, 1910-45.

In 15 of the 36 years from 1910 to 1945, farm operators and unpaid family workers averaged less labor return per hour than did hired farm workers. Returns to farm labor and capital fluctuate with changes in prices occasioned by wars and depressions.

The amount left for the use of land, labor, and capital after deducting all other operating costs from gross income to agriculture has varied during the 36 years, 1910–45, from a low of about 3.5 billion dollars in 1932 to a high of 17 billion dollars in 1945 (section 1, fig. 16). Not only were returns to labor and capital high in 1945, but both cash and total costs were high and farm prosperity was very high, compared with depression and farm foreclosure conditions in 1932 when cash costs, total costs, and returns to labor and capital were very low. Here again, prices of the products sold completely dominated the situation. Efficiencies in production brought about by mechanization and other technological developments could not possibly offset the effects of 40-cent wheat, 6.5-cent cotton, and 3.5-cent hogs.

It appears that, over the third of a century of wars and peace from 1910 to 1945, farmers as a group have earned high wages only in the war periods, after allowances are made for returns to land and farm working capital at prevailing rates. The war years are the outstanding periods in which the farmer and his family received net more per hour for their labor than they paid to hired farm workers (section 2, fig. 16). The same was true during the recovery period following the depression of the 1930's, but in the depression years and during the years before and after World War I, farmers' net return per hour for their labor averaged less than wages paid to hired workers. But it is necessary to recall, here, that farmers in general pay more for land rents and for land, in periods of relative prosperity, and the higher income imputed to land reduces to some extent the returns for their labor.

TABLE 33.—*Annual returns to all farm labor, hired workers, and operator and unpaid family workers, by periods, United States, 1910–44* [1]

Period	Total returns			Returns per worker		
	All farm labor	Hired workers	Operator and family workers	All farm labor	Hired workers	Operator and family workers
	Million dollars	Million dollars	Million dollars	Dollars	Dollars	Dollars
Average of:						
1910–14	3,112	784	2,328	258	271	254
1915–19	5,254	1,139	4,115	451	394	470
1920–24	4,520	1,301	3,219	397	450	379
1925–29	5,269	1,280	3,989	464	433	475
1930–34	2,897	736	2,161	262	287	254
1935–39	4,219	928	3,291	386	362	394
1940–44	8,293	1,557	6,736	803	634	856

[1] Return to operator and family labor is the difference between gross income in agriculture and total production costs exclusive of the imputed costs of operator and family labor. Yearly labor returns per worker were determined by dividing total labor returns to operator and family workers by their average annual employment, and by dividing total wages and perquisites paid to hired labor by hired average annual employment.

It must be remembered that these comparisons are purely relative. They have little bearing on the question of adequate returns to farmers and hired men for their labor. During 1930–34, for example, farmers as a group received only $254 per worker for their year's work; hired men averaged $287 per worker. Living costs were low during these years, but cash was very scarce. In 1940–44, a period of relatively high living costs, labor returns per operator and family worker averaged $856 per year, compared with $634 per hired worker (table 33). It should be noted that these returns to labor do not in the case of operator, family worker, or hired worker, represent cash available for spending. Living supplies furnished by the farm are included in gross income, and the procedure followed allowed the operator a return for the use of his land and working capital before returns to labor were computed. If the farm was owned and free of debt, this allowance would be available for personal expenses or for savings. Members of the family who had jobs off the farm may have added to the funds available for living and savings.

FIGURE 17—PRODUCTION COSTS AND REAL RETURNS TO FARM LABOR, UNITED STATES, 1910-45.
(INDEX NUMBERS, 1935-39=100)

Farmers' gains through more production efficiency have been overshadowed by changes in living costs as well as by changes in prices of items used in farm production. Over the long run there is a fairly close relationship between output per worker and real labor return per worker.

Land rent per acre is inclined to fluctuate with changes in the price level of farm products. In 1932, tenants were paying farm owners a net return per acre of only 3.3 percent on land values. Land values were 107 percent of the 1935–39 average (see table 54 in the appendix). In 1944, on the other hand, farm owners were getting a net return of 9.5 percent on farm real estate that was valued at 138 percent of the

1935–39 average. Net rent per acre in 1932 was 46 percent, and in 1944, 264 percent of the average of the base period.

Different groups among farm workers are affected somewhat differently by price fluctuations. As indicated previously, net hourly returns to labor favor operators and family workers in periods of high prices, and favor hired workers in periods of low prices. Land owners, generally, charge high rents in relation to the value of the farm in periods of prosperity, but tenants are better able to pay high rents then than to pay low rents when prices for farm products are low.

It might be supposed that "real labor returns" [11] per unit of output would be more closely related to the trend in production efficiency than are actual returns to labor. This may be true at certain times, but in general, real labor returns per unit are subject to about the same wide fluctuations as are actual labor returns. Farmers' gains through increased production efficiency, as a group, have been overshadowed by changes in prices of living items as well as by changes in prices of production-cost factors (section 1, fig. 17). During only 7 years out of the 21 years between the wars, 1919–39, were real labor returns per unit of output as good as, or better than, the average return during 1935–39.

The immense increase in output per worker of more than 100 percent during the third of a century, 1910–45, has brought an upward trend in real labor return per worker, especially since the depression of the 1930's and during World War II. This trend has been greatly accelerated at times, and retarded at other times, by violent changes in prices of cost inputs and farm products and in the volume of farm output. These fluctuations stand out prominently in war and depression years (section 2, fig. 17).

The extent to which increases in real labor returns per farm worker have paralleled increases in yearly earnings of those who buy the farm products is another noteworthy question. A comparison of these factors is made here in terms of real yearly income to all farm labor including operator, family, and hired workers, and real wages per year of industrial workers employed in factories, railroading, and mining.

In terms of real dollar income, there is a marked difference during all of the 36-year period between the yearly amounts earned by farm labor and by industrial workers (section 1, fig. 18). It is not a purpose of this publication fully to analyze this difference. It seems sufficient for the purpose intended merely to point out that the dollar earned by the farm worker and the dollar earned by the industrial worker are not comparable because of difference in real purchasing power at any one point of time. A dollar will buy more rent on the farm than it will buy in the city; more food at farm prices than at retail city prices; more heat that comes from farm-produced wood than from coal or oil at city prices. No attempt is made here to indicate what an equitable ratio between dollar income of farm labor and of industrial workers should be. The real significance of the chart for the comparison intended is found in the difference in trends of real labor income for the two classes of workers (section 2, fig. 18).

---

[11] "Real labor returns" were calculated by deflating labor returns in current dollars by the index of prices paid by farmers for commodities used in living. The indexes in figure 17 are for all farm labor, including hired, operator, and family.

This comparison shows that real income per farm worker fluctuates somewhat more than does income per industrial worker. But the labor incomes of the two groups move in the same general direction, and there seems to be no indication that the general trend of one is stronger than the trend of the other. It seems perfectly clear that any benefits

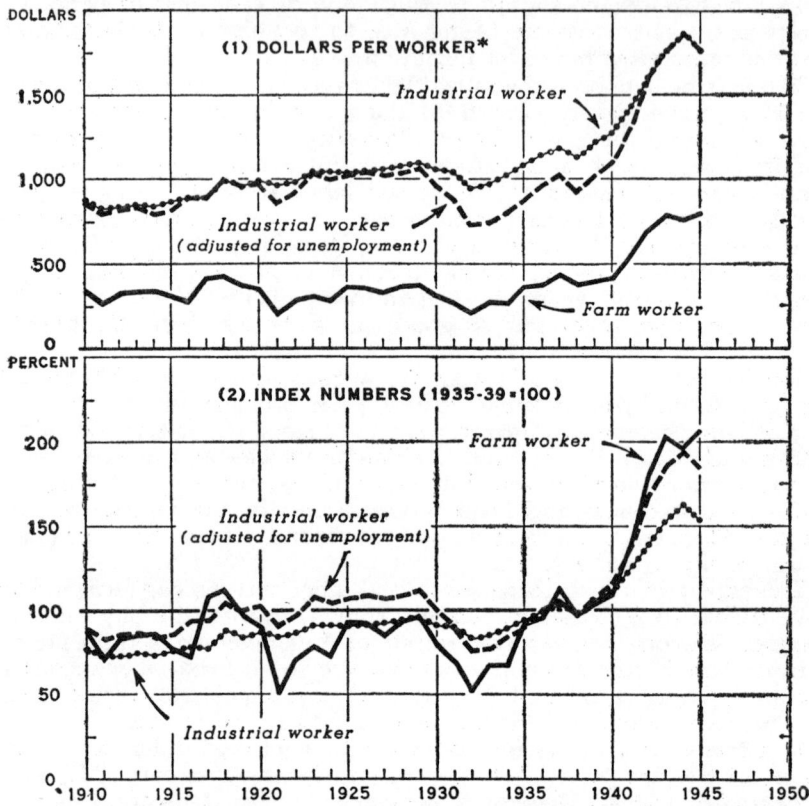

*RETURNS TO LABOR PER FARM WORKER DEFLATED BY INDEX OF PRICES PAID BY FARMERS FOR COMMODITIES USED IN LIVING (1935-39=100), AND AVERAGE ANNUAL WAGES PER INDUSTRIAL WORKER DEFLATED BY B.L.S. "COST-OF-LIVING INDEX (1935-39=100)". ADJUSTMENT OF INDUSTRIAL WORKERS' INCOME FOR UNEMPLOYMENT MADE BY MULTIPLYING PERCENTAGE EMPLOYED IN LABOR FORCE BY UNADJUSTED INCOME PER WORKER.

FIGURE 18—REAL INCOME PER FARM WORKER AND PER INDUSTRIAL WORKER, UNITED STATES, 1910-45.

The marked difference between yearly labor returns of farm workers and of industrial workers is partially offset by cheaper farm items, as food, rent, and fuel. Although real labor returns of farm workers fluctuate more than those of industrial workers, the general trend is about the same for both groups.

received by farm workers in real yearly labor income have been matched by increased real wages of industrial workers. Reductions in production costs of farm products through increased farm efficiency have not increased the returns to agricultural workers any more than wages to industrial workers have been increased. Both groups have been affected by the same set of general economic conditions.

ANOTHER 30 YEARS OF MECHANIZATION

In a way, the most important result of technological developments in agriculture has been the great increase in production of food and fiber for human use. Mechanization was merely an important aid to this end, made possible by a growing Nation that had relatively great resources and dynamic initiative. A growing urban industry that took from the land millions of persons and ever-growing supplies of agricultural products, and at the same time manufactured more and more tractors and other labor-saving machines for farmers, was a motivating influence back of agricultural developments. Agricultural research and education by Federal and State Governments has had great influence. The fact that fewer people were needed in agriculture was merely incidental in the general scheme of development. The fact that industry could get most of its additional workers from the farms was most fortunate in the affairs of a growing Nation.

Two outstanding developments—the switch from animal power to mechanical power, and the increased production per acre and per animal—have made possible large quantities of foods of a wide variety for a growing Nation. At times there appeared to be too much food and fiber because, low priced though it was, people at home and abroad could not buy all they wanted. But at no time, even in the two great war emergencies that have come within the farming experience of many present-day farmers, has there been any real scarcity of either. At no time have our farmers failed to produce abundantly for use at home, and at times to produce something in addition for the hungry people of other Nations.

A recent estimate [12] places the probable population of the United States at 162 million in 1975, compared with an average of 135 million in the war years, 1940–44, and 141 million in 1946. On the basis of this population forecast there will be 21 million more people in our Nation to be fed and clothed, about 30 years from now. If each person is to be fed and clothed at that time as well as the average person was fed and clothed in 1940–44, additional production from an equivalent of 53 million high-yielding crop acres would be necessary. This estimate assumes a continuation of the high production per acre of the war period, and that the same number of acres as were used in 1940–44 (24 million) will be used for products for export (table 34). This does not mean, of course, that additional *new* acreages must be found. Increases in crop yields and in livestock production per animal are probable. Undoubtedly, horses and mules will be much fewer than now, and fewer acres will be needed for their feed, thus releasing cropland for human purposes.

Some idea of the importance of this released acreage for the production of food and fiber can be had by the simple assumption that numbers of horses and mules on farms will continue to decrease at about the same rate as in recent years. On this basis there would be on farms in 1975 only about 4 million head of horses and mules of all ages, compared with about 11 million head in 1946. This would release 27 million average acres between 1940–44 and 1975.

---

[12] Bureau of the Census. Series P-46, No. 7. Population-Special Reports. Sept. 15, 1946

TABLE 34.—*Crop acreages needed to produce food, fiber, and tobacco for domestic consumption in 1955 and 1975, under assumed conditions of mechanization, production, and consumption*

Item	Unit	1940-44 average	Assuming 1935-39 yields and per capita consumption [1]		Assuming 1940-44 yields and per capita consumption [1]	
			1955	1975	1955	1975
Number of horses and mules on farms, January 1 [2]	Millions	13.6	7.5	4.0	7.5	4.0
Number of tractors on farms, January 1 [3]	do	1.9	3.5	5.0	3.5	5.0
Acres of crops harvested for:						
Export products	do	24	24	24	24	24
Feed for horses and mules in cities	do	1	1	1	1	1
Feed for horses and mules on farms	do	38	23	12	21	11
Abnormal military requirements [4]	do	22				
Products for domestic consumption [4]	do	263	310	335	292	316
Total used or needed	do	348	358	372	338	352
Per capita consumption [4]	Acres	1.95	2.07	2.07	1.95	1.95
Production per acre of crops harvested [4]	Index, 1935-39 = 100	113	100	100	113	113
Total population [5]	Millions	135	150	162	150	162
Additional acreage required for domestic consumption	do		47	72	29	53
Acreage released by decline in horse and mule numbers	do		15	26	17	27
Acreage released by elimination of abnormal military requirements	do		22	22	22	22
Net additional acreage needed	do		10	24	−10	4

[1] Acreage required for export products and for feeding horses and mules in cities assumed to remain at 1940-44 average levels.
[2] Horse and mule numbers in 1955 and 1975, assuming a continuation of the recent downward trend in colt crops, and continuation of horse and mule mortality rates of recent years.
[3] Tractor numbers in 1955 and 1975 needed because of the assumed disappearance of work stock (horses and mules over 3 years old). Average displacement of around 4 head of work stock per tractor was assumed to average around 3 between 1944 and 1955, and around 2 between 1955 and 1975. Estimates of tractors in 1955 and 1975 do not include allowance for expected further increase in garden tractors on farms. There were only about 50 thousand of them in 1944.

⁴ So far as can be determined our civilian population consumed an average of 7 to 8 percent more farm products during 1940–44 than during 1935–39. This greater per capita consumption came from fewer acres per capita because of an increase in yields per acre of 13 percent. In addition, the military forces took the same quantity per capita plus the products from about 22 million acres of land. (See discussion begininng on p. 25). The average per capita consumption of 1.95 acres in 1940–44 multiplied by the total population (including military) of 135 million equals 263 million acres. The domestic consumption of 263 million acres in 1940–44 thus includes civilian consumption and military personnel consumption at civilian rates.

⁵ Population year beginning July 1. Forecasts for 1955 and 1975 taken from Bureau of the Census, Series P-46, No. 7, Population-Special Reports, Sept. 15, 1946.

By 1975 an additional 22 million crop acres should be available that were used in 1940–44 for the production of military supplies over and above the high per capita rate of consumption of our civilian and military population. (See footnote 4, table 34). Production of 49 million acres from these two sources would nearly take care of food and fiber requirements for the increased population, at the high consumption rates of 1940–44. Additional production equivalent to that from only 4 million acres would be needed. There are so many possibilities in such a situation that 4 million acres, more or less, is not a significant part of the total. The range in the total crop acreage harvested from time to time is much greater than this. Additional drainage of lands now being farmed could more than make up for this difference. A little better farming all around could easily make up for it and accelerated adoption of improved practices would throw the balance in the other direction.

The important question is whether farmers will be able during the next farming generation to produce at the high wartime level and yet conserve their lands so that 30 years hence the acreage in crops will be as large and as productive as in 1940–44. In the opinion of the authors this will not be difficult. Continuation of a national policy of soil conservation and better farming methods, with technological progress that leads to more production per acre and per animal, may be expected. Increased adoption of improved production techniques could lead to surplus production, requiring expanded markets for disposal. Experience of the last 25 years points strongly in this direction.

With the above assumptions of 1940–44 average production per acre and per animal, and even with a continuation of the high wartime level of per capita consumption, farm surpluses are likely to pile up long before 1975, unless markets are expanded. By 1955, about 17 million crop acres that were used to produce horse and mule feed during the war would be released for other production. This acreage, plus the 22 million acres that were used to produce extra war supplies, would make an additional 39 million crop acres available for production for human use, whereas only 29 million acres will be required under the assumptions used for the expected increase in population (table 34).

There are, of course, two sides to all questions. If it is assumed that farmers' average level of production per acre and per animal would be like that of 1935–39, instead of the high level of 1940–44, 72 million additional crop acres would be needed in 1975 to supply each person with the average per capita rate of consumption that prevailed in 1935–39. The assumed disappearance of horses and mules would make available 26 million acres and this, plus the 22 million acres used to produce extra war supplies, would account for 48 of the required 72 million acres. Production equivalent to that of 24 million average acres in 1935–39 would have to come from other sources. Therefore the problem under these assumed conditions would be one of increased production rather than one of expanded markets.

If the need for more farm products develops, however, several ways will be available to farmers for increasing production without more land, as witnessed by their performance in World War II. Under assumptions of prewar yields and per capita consumption, additional production from 10 million 1935–39 acres would be needed by the increased population about 1955.

These assumptions and observations apply to agriculture in general. Production of many crops in the United States can be expanded tremendously by the use of small acreages. Fruit, vineyard, and nut crops as a whole used less than 5 million acres in 1944. All truck crops for processing and fresh consumption used a little more than 4 million acres. Irish potatoes were grown on only 3 million acres in 1944. Thus, production of these important crops took only about 12 million acres, or less than 5 percent of the total crop acreage in 1944. Production of these crops can be expanded greatly by using relatively little additional land.

Any increased domestic needs for such cash crops as cotton and wheat could be met by using lands suitable for their growth. Increased cotton acreages could come largely from acres now used to grow feed crops, and wheat-acreage expansion could be mainly on lands now used for pasture and feed crops. The tremendous wartime expansion of soybeans and peanuts was principally at the expense of feed crops in the Corn Belt in the first instance, and at the expense of feed crops and cotton in the second instance. It probably will be easy at any time in the next 30 years to divert enough suitable land to any of the chief cash-crop enterprises to over-supply the domestic markets. Once our requirements for direct food, fiber, and tobacco crops are met, the remainder of our agricultural resources would become available for feed production to be converted into meat animals and animal products. Great possibilities are ahead for increasing feed production through improvement of pastures and hay crops, and for increased production of some grain crops, especially in parts of the South.

Assumptions regarding future horse and mule numbers have a decided effect on forecasts of farm mechanization. In recent years, one farm tractor has been added to the numbers on farms for approximately each 4 head of work stock that disappear. This ratio of work-stock disappearance to tractors added is bound to decrease as tractors and tractor equipment for small farms become increasingly available, and as thousands of farms acquire one or more additional tractors. For the purpose of the estimates used in this section it has been assumed that during the 10 years, 1945–55, the net disappearance of work stock would average 3 per additional tractor, and during the 20 years, 1955–75, one tractor would be added for each two work animals that disappear.

On the basis of the above assumptions, and the assumption that horse and mule numbers will continue to disappear from farms at about the same rate as in recent years, the tractors on farms in 1955 would number about 3.5 million, and in 1975, about 5.0 million. Suitable equipment for these power units will be developed, so that farming will become increasingly attractive from the standpoint of lessened physical effort.

Mechanical cotton pickers are a present-day development of keen interest. The next generation may see their successful development, and witness widespread adoption in some cotton areas. Many people believe that we are on the verge of rapid development of farm mechanization in southern agriculture. The effects of such a development on farm workers and on the pattern of agriculture during the next generation may be enormous. In other parts of the country such recently developed machines as beet harvesters, sweet-corn pickers, pick-up balers, forage choppers, manure loaders, dairy-barn cleaners, flame cultivators, and machines for removing silage, as well as further improvement and more widespread adoption of older labor-saving machines are now on the horizon.

Further farm mechanization in the United States will mean that fewer people will be needed for farm work, but of equal importance is the possibility of less need for overworking at rush seasons and for women and children working in the fields. Something resembling complete electric service for farmers seems assured and this will help to reduce the long hours and tiresome jobs in the homes as well as on the farms.

Obviously, the conclusions reached in this section of this publication are contingent to a large extent on the assumptions covering further disappearance of horses and mules from farms. About the best that can be said of these assumptions is that they seem reasonable and logical. If horse and mule numbers are reduced slower or faster than here indicated, the periods indicated by the dates used will be lengthened or shortened. The stage is definitely set for continued mechanization. New labor-saving power units and machines will be adopted most rapidly if we have satisfactory economic conditions. New machines and new ideas will be brought out for trial, and many of these will be adopted during the coming farming generation, just as the past generation adopted new practices and new techniques.

The evidence developed in this publication seems to point conclusively to the fact that a very large part of the increased efficiency in agricultural production in the past, or the increase in output per unit of input, has been caused principally by increased production per acre and per animal rather than by decreased total physical expenditures. Initial savings in labor and other costs of producing farm-animal power have been used to produce more milk, more pork, and more of other livestock products for the market. Labor saved because of greater speed in doing farm jobs by the use of modern machines has not always meant the release of farm workers. Rather, the large volume of business developed through the introduction of more intensive enterprises, and through more production per acre and per animal, has absorbed much of the labor that would have been released because of mechanization.

Further increases in efficiency of farm production are desirable and will continue. If past patterns continue in the future march of farm technology, increases in efficiency will be closely related to increases in total volume of farm output.

Production efficiency and the relationships between prices that farmers receive and the prices they pay will continue to be dominant factors in determining the economic welfare of farm people. Expanded markets for farm products must accompany increases in production efficiency if both producers and consumers are to benefit to the fullest extent.

APPENDIX

The following tables contain data used in the charts shown in this report, and other supplementary information.

PROGRESS OF FARM MECHANIZATION

TABLE 35.—*Index numbers of total volume of farm power and machinery and equipment, farm output, and total farm employment, United States, 1870–1946* [1] [2]

[Volume in terms of 1935–39 average dollars; 1870 = 100]

Year	Farm power machinery and equipment	Farm output	Farm employment	Year	Farm power machinery and equipment	Farm output	Farm employment
1870	100	100	100	1927	463	326	157
1880	163	156	125	1928	463	337	158
1890	232	183	146	1929	466	334	158
1900	295	240	159	1930	471	324	156
1910	385	260	169	1931	468	355	156
1911	395	271	168	1932	452	347	154
1912	404	274	168	1933	416	318	154
1913	416	279	168	1934	391	269	151
1914	429	284	168	1935	389	326	155
1915	430	303	167	1936	391	292	154
1916	431	277	168	1937	403	371	152
1917	420	269	165	1938	419	360	151
1918	422	290	158	1939	428	365	150
1919	468	295	155	1940	434	373	148
1920	477	313	159	1941	445	389	145
1921	505	279	159	1942	466	433	145
1922	496	305	160	1943	487	425	143
1923	454	308	159	1944	494	446	140
1924	455	308	159	1945	517	440	137
1925	458	318	160	1946	542	453	140
1926	460	326	161				

[1] See footnotes table 4 for sources of data.
[2] Data used in figure 1 derived from this table.

TABLE 36.—*Index numbers of production per worker in agriculture and industry, United States, 1910–45* [1]

[1935–39 = 100]

Year	Gross farm production per farm worker	Production per worker in manufacturing and mining	Year	Gross farm production per farm worker	Production per worker in manufacturing and mining
1910	77	62	1929	98	101
1911	80	59			
1912	82	66	1930	96	95
1913	83	68	1931	103	92
1914	85	64	1932	101	85
1915	89	69	1933	94	91
1916	84	69	1934	83	85
1917	83	66	1935	95	93
1918	91	64	1936	86	102
1919	94	65	1937	107	102
			1938	105	95
1920	97	68	1939	107	108
1921	87	67			
1922	92	78	1940	110	115
1923	94	81	1941	117	122
1924	93	81	1942	128	130
1925	94	87	1943	127	138
1926	95	90	1944	135	143
1927	97	92	1945	137	142
1928	99	96			

[1] Data used in figure 2 derived from this table.

TABLE 37.—*Index numbers of animal units of breeding livestock and livestock production per breeding unit, United States, 1919–46* [1]

[1935–39 = 100]

Year	Animal units of breeding livestock	Livestock production per breeding unit	Year	Animal units of breeding livestock	Livestock production per breeding unit
1919....	105	80	1933...	112	95
			1934...	110	87
1920....	102	80	1935...	97	95
1921....	102	83	1936...	101	98
1922....	106	87	1937...	99	98
1923....	110	86	1938...	98	104
1924....	106	88	1939...	105	105
1925....	101	91			
1926....	100	95	1940...	108	103
1927....	103	95	1941...	107	110
1928....	102	96	1942...	117	111
1929....	101	98	1943...	130	109
			1944...	129	105
1930...	102	99	1945...	120	112
1931...	104	99	1946...	119	111
1932....	107	97			

[1] Includes all breeding livestock except horses, and all livestock production except farm-produced power of horses and mules.

TABLE 38.—*Index numbers of horse and mule numbers on farms, and volume of production of meat animals and animal products, United States, 1919–46* [1]

[1920–24 = 100]

Year	Horses and mules all ages	Product added by meat animals and animal products	Year	Horses and mules all ages	Product added by meat animals and animal products
1919....	108	95	1933...	71	118
			1934...	69	111
1920....	105	92	1935...	68	108
1921....	102	97	1936...	66	114
1922....	100	102	1937...	64	113
1923....	98	105	1938...	62	116
1924....	95	104	1939...	60	124
1925....	92	105			
1926....	90	107	1940...	59	125
1927....	86	109	1941...	57	133
1928....	83	110	1942...	56	146
1929....	80	111	1943...	54	157
			1944...	51	155
1930....	78	114	1945...	49	156
1931....	75	116	1946...	45	149
1932....	73	116			

[1] Data used in figure 3 derived from this table.

TABLE 39.—*Inventory values of horses and mules and farm machinery, United States, January 1, 1910–46* [1]

[Values in current dollars]

Year	Horses and mules [2]	Tractors	Motor-trucks	Auto-mobiles	Other farm machinery	Total
	Million dollars	*Million dollars*	*Million dollars*	*Million dollars*	*Million dollars*	*Million dollars*
1910	2,790	2	0	25	1,238	4,055
1911	2,957	6	1	53	1,261	4,278
1912	2,872	10	3	85	1,284	4,254
1913	3,039	16	4	111	1,328	4,498
1914	3,080	21	5	133	1,369	4,608
1915	2,936	30	10	189	1,377	4,542
1916	2,912	42	18	291	1,392	4,655
1917	2,985	49	23	356	1,424	4,837
1918	3,104	66	31	520	1,663	5,384
1919	3,043	180	43	628	2,151	6,045
1920	3,072	283	67	1,005	2,240	6,667
1921	2,553	358	103	1,009	2,423	6,446
1922	2,051	347	92	687	2,078	5,255
1923	1,962	263	96	654	1,671	4,646
1924	1,814	285	101	751	1,680	4,631
1925	1,724	287	115	705	1,663	4,494
1926	1,683	330	139	791	1,612	4,555
1927	1,552	362	171	855	1,617	4,557
1928	1,565	403	186	783	1,628	4,565
1929	1,575	416	222	857	1,621	4,691
1930	1,539	461	244	963	1,634	4,841
1931	1,276	488	213	829	1,615	4,421
1932	1,078	437	190	756	1,479	3,940
1933	1,038	360	159	642	1,235	3,434
1934	1,269	296	152	600	1,067	3,384
1935	1,451	294	160	698	1,001	3,604
1936	1,736	361	164	758	1,023	4,042
1937	1,762	452	186	814	1,107	4,321
1938	1,581	568	220	906	1,262	4,537
1939	1,446	604	230	924	1,367	4,571
1940	1,328	589	238	944	1,364	4,463
1941	1,170	636	257	912	1,421	4,396
1942	1,102	808	300	1,015	1,625	4,850
1943	1,295	1,038	425	987	2,059	5,804
1944	1,286	1,152	570	1,091	2,566	6,665
1945	1,078	1,330	631	1,088	3,186	7,313
1946	942	1,316	586	922	3,433	7,199

[1] Data used in figure 4 derived from this table.
[2] Includes harness.

TABLE 40.—*Inventory values of farm horses and mules, and farm machinery, United States, January 1, 1910–46* [1]

[Values in 1935–39 average dollars]

Year	Horses and mules [2]	Tractors	Motor-trucks	Automobiles	Other farm machinery	Total
	Million dollars	*Million dollars*	*Million dollars*	*Million dollars*	*Million dollars*	*Million dollars*
1910	2,564	1	0	13	1,876	4,454
1911	2,631	2	1	26	1,911	4,571
1912	2,679	4	1	46	1,945	4,675
1913	2,719	7	3	67	2,012	4,808
1914	2,787	8	4	89	2,074	4,962
1915	2,813	12	7	123	2,025	4,980
1916	2,787	18	12	179	1,989	4,985
1917	2,782	26	17	251	1,780	4,856
1918	2,788	42	26	391	1,630	4,877
1919	2,797	79	32	458	2,049	5,415
1920	2,748	123	40	558	2,055	5,524
1921	2,689	172	60	619	2,308	5,848
1922	2,618	186	76	630	2,234	5,744
1923	2,542	214	92	681	1,723	5,252
1924	2,470	248	105	781	1,663	5,267
1925	2,386	274	133	854	1,647	5,294
1926	2,315	310	162	937	1,596	5,320
1927	2,228	346	192	993	1,601	5,360
1928	2,147	391	218	993	1,612	5,361
1929	2,079	414	244	1,032	1,621	5,390
1930	2,019	460	261	1,075	1,634	5,449
1931	1,953	498	267	1,060	1,631	5,409
1932	1,873	511	264	987	1,590	5,225
1933	1,807	510	251	884	1,357	4,809
1934	1,756	508	254	884	1,123	4,525
1935	1,728	524	258	947	1,043	4,500
1936	1,680	562	268	971	1,044	4,525
1937	1,636	616	287	1,030	1,096	4,665
1938	1,579	684	302	1,068	1,214	4,847
1939	1,536	724	296	1,048	1,353	4,957
1940	1,503	772	304	1,077	1,364	5,020
1941	1,464	838	318	1,089	1,434	5,143
1942	1,417	945	336	1,114	1,578	5,390
1943	1,378	1,050	371	1,086	1,746	5,631
1944	1,319	1,105	397	1,071	1,820	5,712
1945	1,257	1,212	423	1,066	2,025	5,983
1946	1,168	1,292	450	1,066	2,296	6,272

[1] Data used in figure 5 derived from this table.
[2] Includes harness.

TABLE 41.—*Horses and mules, and tractors on farms, United States, January 1, 1910–47* [1]

Year	Horses and mules	Tractors	Year	Horses and mules	Tractors
	Thousands	Thousands		Thousands	Thousands
1910	24,211	1	1930	19,124	920
1911	24,847	4	1931	18,468	997
1912	25,277	8	1932	17,812	1,022
1913	25,691	14	1933	17,337	1,019
1914	26,178	17	1934	16,997	1,016
1915	26,493	25	1935	16,683	1,048
1916	26,534	37	1936	16,226	1,125
1917	26,659	51	1937	15,802	1,230
1918	26,723	85	1938	15,245	1,370
1919	26,490	158	1939	14,792	1,445
1920	25,742	246	1940	14,478	1,545
1921	25,137	343	1941	14,104	1,675
1922	24,588	372	1942	13,655	1,890
1923	24,018	428	1943	13,231	2,100
1924	23,285	496	1944	12,613	2,210
1925	22,569	549	1945	11,950	2,425
1926	21,986	621	1946	11,063	2,585
1927	21,192	693	1947	10,024	2,700
1928	20,448	782			
1929	19,744	827			

[1] Data used in figure 6 derived from this table.

TABLE 42.—*Tractors: Number on farms by geographic divisions and United States, 1920, 1930, 1940, and 1945* [1] [2]

Year	New England	Middle Atlantic	East North Central	West North Central	South Atlantic	East South Central	West South Central	Mountain	Pacific	United States
	Thousands	Thousands	Thousands	Thousands	Thousands	Thousands	Thousands	Thousands	Thousands	Thousands
1920	2	14	58	98	11	5	20	18	20	246
1930	14	82	249	318	47	25	74	48	63	920
1940	28	125	431	531	62	41	163	74	90	1,545
1945	54	206	644	759	135	87	277	120	140	2,422

[1] From reports of the Census of Agriculture.
[2] Data used in figure 7 derived from this table.

TABLE 43.—*Breaking land:* [1] *Percentage broken with tractor-drawn implements, by States, 1939* [2]

State	Acreage broken with tractor power	State	Acreage broken with tractor power
	Percent		*Percent*
Maine	28	North Carolina	15
New Hampshire	47	South Carolina	14
Vermont	26	Georgia	7
Massachusetts	51	Florida	19
Rhode Island	60		
Connecticut	50	Kentucky	17
		Tennessee	13
New York	54	Alabama	10
New Jersey	69	Mississippi	17
Pennsylvania	43		
		Arkansas	17
Ohio	63	Louisiana	38
Indiana	69	Oklahoma	70
Illinois	80	Texas	72
Michigan	53		
Wisconsin	56	Montana	84
		Idaho	51
Minnesota	72	Wyoming	72
Iowa	82	Colorado	75
Missouri	50	New Mexico	64
North Dakota	79	Arizona	88
South Dakota	80	Utah	50
Nebraska	76	Nevada	61
Kansas	88		
Delaware	55	Washington	78
Maryland	36	Oregon	68
Virginia	15	California	88
West Virginia	11		

[1] Breaking land includes plowing, listing, bedding, and middlebusting. State averages were computed from county data used in making figure 8.
[2] A. P. Brodell. Machine and Hand Methods in Crop Production. F.M. 18, p. 6. Revised January 1942. Processed.

TABLE 44.—*Value of horses and mules, tractors, motortrucks, automobiles, and other farm machinery, per acre of cropland, by geographic divisions, 1910, 1920, 1930, 1940–45* [1]

[Values in 1935–39 average dollars]

NEW ENGLAND

Item	1910	1920	1930	1940	1941	1942	1943	1944	1945
	Dollars	Dollars	Dollars	Dollars	Dollars	Dollars	Dollars	Dollars	Dollars
Other machinery..	15.85	13.32	14.40	8.85	9.42	10.23	11.49	11.81	13.46
Automobiles.....	.05	2.60	6.03	7.28	7.29	7.43	7.34	7.14	9.14
Motortrucks.....		.51	2.87	4.06	4.21	4.40	4.91	5.27	5.96
Tractors.........		.17	1.43	3.14	3.35	3.75	4.30	4.51	5.06
Horses and mules.	12.00	11.04	8.38	5.94	5.73	5.44	5.45	5.29	5.13
Total.....	27.90	27.64	33.11	29.27	30.00	31.25	33.49	34.02	38.75
	MIDDLE ATLANTIC									
Other machinery..	13.64	13.17	14.73	11.43	12.00	13.21	14.95	14.59	17.00
Automobiles.....	.05	2.40	5.68	6.04	6.10	6.25	6.22	5.74	6.36
Motortrucks.....		.35	2.43	2.45	2.56	2.72	3.06	3.08	3.12
Tractors.........		.34	2.50	4.00	4.33	4.74	5.49	5.45	6.26
Horses and mules.	10.47	9.98	7.82	6.78	6.64	6.39	6.28	5.61	5.56
Total......	24.16	26.24	33.16	30.70	31.63	33.31	36.00	34.47	38.30
	EAST NORTH CENTRAL									
Other machinery..	6.36	6.44	4.33	5.39	5.61	6.09	6.64	6.78	7.39
Automobiles.....	.05	2.21	3.93	4.52	4.52	4.55	4.38	4.23	3.99
Motortrucks.....		.12	.99	.98	1.01	1.06	1.15	1.21	1.13
Tractors.........		.41	2.00	3.56	3.80	4.23	4.64	4.75	5.10
Horses and mules.	8.08	7.57	5.65	4.54	4.26	3.91	3.61	3.25	3.01
Total......	14.49	16.75	16.90	18.99	19.20	19.84	20.42	20.22	20.62
	WEST NORTH CENTRAL									
Other machinery..	4.56	4.87	4.29	2.98	3.12	3.42	3.62	3.79	4.24
Automobiles.....	.04	1.36	1.92	2.03	2.05	2.10	1.96	1.94	1.83
Motortrucks.....		.07	.36	.43	.45	.48	.50	.54	.60
Tractors.........		.38	1.16	2.11	2.28	2.55	2.69	2.83	3.12
Horses and mules.	5.51	5.32	3.62	2.54	2.49	2.41	2.24	2.11	2.01
Total......	10.11	12.00	11.35	10.09	10.39	10.96	11.01	11.21	11.80
	SOUTH ATLANTIC									
Other machinery..	4.92	5.10	1.60	1.97	2.13	2.32	2.55	2.73	3.12
Automobiles.....	.04	1.68	4.46	4.21	4.37	4.39	4.25	4.32	5.13
Motortrucks.....		.15	.98	1.17	1.25	1.30	1.43	1.57	1.87
Tractors.........		.15	.71	.91	1.01	1.16	1.29	1.44	1.63
Horses and mules.	9.09	10.00	8.33	7.34	7.44	7.25	7.06	7.15	6.97
Total......	14.05	17.08	16.08	15.60	16.20	16.42	16.58	17.21	18.72
	EAST SOUTH CENTRAL									
Other machinery..	4.24	4.46	2.15	2.59	2.75	2.97	3.24	3.52	4.01
Automobiles.....	.02	.85	3.36	2.87	2.93	2.92	2.81	2.88	3.19
Motortrucks.....		.06	.50	.80	.84	.87	.95	1.05	1.21
Tractors.........		.09	.45	.74	.82	.92	1.01	1.14	1.28
Horses and mules.	9.12	10.15	8.22	7.55	7.51	7.36	7.13	7.31	7.08
Total......	13.38	15.61	14.68	14.55	14.85	15.04	15.14	15.90	16.77

TABLE 44.—*Value of horses and mules, tractors, motortrucks, automobiles, and other farm machinery, per acre of cropland, by geographic divisions, 1910, 1920, 1930, 1940–45* [1]—Continued

[Values in 1935–39 average dollars]

WEST SOUTH CENTRAL

Item	1910	1920	1930	1940	1941	1942	1943	1944	1945
	Dollars	Dollars	Dollars	Dollars	Dollars	Dollars	Dollars	Dollars	Dollars
Other machinery..	4.26	3.53	2.46	2.44	2.61	2.94	3.23	3.34	4.01
Automobiles.....	.03	.93	2.32	2.11	2.18	2.28	2.21	2.17	2.25
Motortrucks.....		.05	.47	.61	.65	.70	.77	.82	1.04
Tractors.........		.19	.62	1.41	1.56	1.87	2.09	2.19	2.58
Horses and mules.	7.23	6.42	4.90	3.56	3.52	3.44	3.33	3.16	3.22
Total......	11.52	11.12	10.77	10.13	10.52	11.23	11.63	11.68	13.10

MOUNTAIN

Other machinery..	6.60	4.86	4.56	3.58	3.61	3.94	4.22	4.43	5.12
Automobiles.....	.05	1.15	1.72	1.82	1.75	1.78	1.68	1.66	1.66
Motortrucks.....		.10	.59	.84	.84	.88	.94	1.01	1.26
Tractors.........		.44	.97	1.58	1.65	1.90	2.02	2.16	2.45
Horses and mules.	8.45	6.33	3.87	2.74	2.59	2.53	2.44	2.35	2.31
Total......	15.10	12.88	11.71	10.56	10.44	11.03	11.30	11.61	12.80

PACIFIC

Other machinery..	5.96	7.44	6.37	5.91	6.29	6.71	7.35	7.50	8.37
Automobiles.....	.05	1.83	3.44	3.91	3.98	3.95	3.81	3.68	3.90
Motortrucks.....		.20	1.10	1.60	1.70	1.74	1.90	1.99	2.28
Tractors.........		.69	2.13	2.79	3.18	3.36	3.66	3.82	4.20
Horses and mules.	6.21	5.86	3.47	2.43	2.36	2.21	2.10	1.90	1.80
Total......	12.22	16.02	16.51	16.64	17.51	17.97	18.82	18.89	20.55

UNITED STATES

Other machinery..	5.67	5.58	4.28	3.73	3.92	4.30	4.75	4.84	5.48
Automobiles.....	.04	1.51	2.82	2.94	2.98	3.03	2.96	2.85	2.93
Motortrucks.....		.11	.68	.83	.87	.92	1.01	1.06	1.17
Tractors.........		.33	1.20	2.11	2.29	2.57	2.86	2.94	3.28
Horses and mules.	7.33	6.99	5.01	3.95	3.86	3.72	3.60	3.36	3.24
Total......	13.04	14.52	13.99	13.56	13.92	14.54	15.18	15.05	16.10

[1] Based on data from reports of the Census of Agriculture.

TABLE 45.—*Value of products produced on farms reporting different types of power, by geographic divisions, 1939* [1] [2]

Geographic division	Total value of products	Percentage of total value on farms reporting		
		Tractors	Horses and/or mules only	No horses, mules, or tractors
	Million dollars	*Percent*	*Percent*	*Percent*
New England	236.2	41.0	36.4	22.6
Middle Atlantic	587.2	56.4	33.0	10.6
East North Central	1,483.8	65.7	28.7	5.6
West North Central	1,830.7	68.5	27.0	4.5
South Atlantic	915.8	21.1	64.1	14.8
East South Central	609.7	15.5	66.0	18.5
West South Central	959.6	39.3	51.4	9.3
Mountain	486.8	50.6	43.0	6.4
Pacific	703.8	57.0	23.2	19.8
United States	7,813.6	50.9	39.0	10.1

[1] See footnote 1, table 26.
[2] Data used in figure 9 derived from this table.

TABLE 46.—*Percentage of all farms receiving central-station electric service, United States, July 1, 1945* [1]

State	Percentage of farms receiving electric service	State	Percentage of farms receiving electric service
	Percent		*Percent*
Maine	56.1	West Virginia	33.3
New Hampshire	73.7	North Carolina	37.0
Vermont	59.4	South Carolina	37.3
Massachusetts	60.9	Georgia	34.0
Rhode Island	81.8	Florida	35.6
Connecticut	84.4		
		Kentucky	27.0
New York	81.0	Tennessee	27.1
New Jersey	88.8	Alabama	27.2
Pennsylvania	66.4	Mississippi	20.6
Ohio	83.0	Arkansas	20.4
Indiana	76.3	Louisiana	23.9
Illinois	60.1	Oklahoma	21.3
Michigan	88.9	Texas	36.7
Wisconsin	69.8		
		Montana	30.1
Minnesota	48.9	Idaho	82.2
Iowa	61.5	Wyoming	42.4
Missouri	30.0	Colorado	52.8
North Dakota	9.0	New Mexico	24.9
South Dakota	12.1	Arizona	62.9
Nebraska	33.2	Utah	76.0
Kansas	30.8	Nevada	53.7
Delaware	56.6	Washington	82.9
Maryland	63.6	Oregon	77.8
Virginia	34.3	California	87.7

[1] Based on data in reports of the Rural Electrification Administration, United States Department of Agriculture.

TABLE 47.—*Total production costs of farm power, machinery, and labor, United States, 1910–45* [1]

[Costs in 1935–39 average dollars]

Year	Total farm-produced power	Tractors	Trucks and automobiles [2]	Other farm machinery	Farm labor [3]	Total
	Million dollars	*Million dollars*	*Million dollars*	*Million dollars*	*Million dollars*	*Million dollars*
1910	1,535	2	10	781	2,999	5,327
1911	1,584	5	21	796	3,048	5,454
1912	1,623	8	30	821	3,074	5,556
1913	1,662	10	41	834	3,073	5,620
1914	1,705	13	53	844	3,114	5,729
1915	1,735	22	80	818	3,190	5,845
1916	1,752	32	127	788	3,057	5,756
1917	1,768	50	173	726	3,063	5,780
1918	1,687	70	199	670	3,154	5,780
1919	1,781	112	231	828	3,145	6,097
1920	1,727	155	259	832	3,233	6,206
1921	1,681	145	302	852	3,023	6,003
1922	1,638	171	315	830	3,096	6,050
1923	1,596	163	357	679	3,135	5,930
1924	1,544	179	417	656	3,078	5,874
1925	1,491	186	463	649	3,119	5,908
1926	1,448	211	520	635	3,138	5,952
1927	1,394	236	559	637	3,132	5,958
1928	1,342	254	589	646	3,162	5,993
1929	1,293	286	626	654	3,151	6,010
1930	1,251	310	635	649	3,101	5,946
1931	1,207	297	589	617	3,163	5,873
1932	1,163	269	542	589	3,136	5,699
1933	1,132	258	508	511	3,059	5,468
1934	1,109	256	512	440	2,857	5,174
1935	1,091	294	550	443	3,042	5,420
1936	1,063	323	586	459	2,921	5,352
1937	1,040	381	600	495	3,083	5,599
1938	1,006	412	591	528	3,006	5,543
1939	978	455	628	570	3,022	5,653
1940	956	509	623	575	3,015	5,678
1941	933	571	647	621	3,061	5,833
1942	905	638	676	689	3,155	6,063
1943	872	677	697	747	3,126	6,119
1944	832	751	718	807	3,161	6,269
1945	789	807	747	897	3,070	6,310

[1] Data for figure 10 derived from this table.
[2] Includes only farm production share of automobile costs.
[3] Includes all farm labor costs except those for servicing and maintenance of machinery, trucks, automobiles, and tractors, and for the care and maintenance of and the growing of feed for horses and mules, which are included in the power and machinery cost items.

TABLE 48.—*Total production costs of farm power, machinery, and labor, United States, 1910–45* [1]

[Costs in current dollars]

Year	Total farm-produced power	Tractors	Trucks and automobiles [2]	Other farm machinery	Farm labor [3]	Total
	Million dollars	*Million dollars*	*Million dollars*	*Million dollars*	*Million dollars*	*Million dollars*
1910	1,554	4	17	556	2,545	4,676
1911	1,571	7	28	565	2,561	4,732
1912	1,893	13	40	583	2,653	5,182
1913	1,648	15	49	592	2,697	5,001
1914	1,791	20	64	599	2,654	5,128
1915	1,816	29	92	595	2,705	5,237
1916	1,806	39	137	603	2,923	5,508
1917	2,297	62	203	650	3,666	6,878
1918	2,866	96	270	756	4,579	8,567
1919	3,361	162	343	971	5,281	10,118
1920	3,441	230	438	1,036	6,165	11,310
1921	2,254	199	403	941	3,657	7,454
1922	1,715	194	367	826	3,674	6,776
1923	1,831	180	388	708	4,072	7,179
1924	1,893	191	437	702	4,084	7,307
1925	1,958	207	495	702	4,168	7,530
1926	1,721	240	569	690	4,278	7,498
1927	1,697	256	582	687	4,169	7,391
1928	1,575	273	610	694	4,160	7,312
1929	1,522	304	654	696	4,170	7,346
1930	1,428	323	638	681	3,790	6,860
1931	1,174	283	531	614	2,984	5,586
1932	834	246	474	535	2,197	4,286
1933	675	223	438	451	1,998	3,785
1934	921	228	466	413	2,275	4,303
1935	1,279	268	504	423	2,647	5,121
1936	934	304	547	449	2,820	5,054
1937	1,265	378	586	503	3,304	6,036
1938	925	410	594	546	3,165	5,640
1939	795	437	625	576	3,133	5,566
1940	858	465	618	576	3,203	5,720
1941	869	538	670	639	3,898	6,614
1942	1,046	644	766	751	5,146	8,353
1943	1,200	762	878	937	6,668	10,445
1944	1,441	905	1,000	1,166	7,861	12,373
1945	1,438	999	1,070	1,421	8,478	13,406

[1] Data for figure 11 derived from this table.
[2] Includes only farm production share of automobile costs.
[3] Includes all farm labor costs except those for servicing and maintenance of machinery, trucks, automobiles, and tractors, and for the care and maintenance of and the growing of feed for horses and mules, which are included in the power and machinery cost items.

TABLE 49.—*Index numbers of total production costs and costs per unit of farm output for farm power, machinery, and labor, United States, 1910–45* [1] *(costs in 1935–39 average dollars)*

[1935–39 = 100]

Year	Farm-produced power		Tractors		Trucks and automobiles [2]		Other machinery		All power and machinery		All farm labor, power, and machinery	
	Total cost	Cost per unit of farm output	Total cost	Cost per unit of farm output	Total cost	Cost per unit of farm output	Total cost	Cost per unit of farm output	Total cost	Cost per unit of farm output	Total cost	Cost per unit of farm output
1910	148	195	1	1	2	3	157	207	93	122	97	128
1911	153	194	1	1	4	5	160	203	96	122	99	125
1912	157	196	2	2	5	6	165	206	99	124	101	126
1913	161	199	3	4	7	9	167	206	102	126	102	126
1914	165	199	3	4	9	11	169	204	105	127	104	125
1915	168	191	6	7	14	16	164	186	106	120	106	120
1916	169	209	9	11	21	26	158	195	108	133	104	128
1917	171	219	13	17	29	37	145	186	109	140	105	135
1918	163	194	19	23	34	40	134	160	105	125	105	125
1919	172	202	30	35	39	46	166	195	118	139	111	131
1920	167	182	42	46	44	48	167	182	119	129	113	123
1921	162	200	39	48	51	63	171	211	119	147	109	135
1922	158	178	46	52	53	60	166	187	118	133	110	124
1923	154	171	44	49	60	67	136	151	112	124	108	120
1924	149	166	48	53	71	79	131	146	112	124	107	119
1925	144	155	50	54	78	84	130	140	112	120	107	115
1926	140	147	57	60	88	93	127	134	113	119	108	114
1927	135	142	63	66	95	100	128	135	113	119	108	114
1928	130	131	68	69	100	101	129	130	113	114	109	110
1929	125	129	77	79	106	109	131	135	114	118	109	112

Year												
1930	121	127	83	87	107	113	130	137	114	120	108	114
1931	117	112	80	77	100	96	124	119	108	104	107	103
1932	112	111	72	71	92	91	118	117	103	102	103	102
1933	109	117	69	74	86	92	102	110	96	103	99	106
1934	107	135	69	87	87	110	88	111	93	118	94	119
1935	105	108	79	83	93	96	89	92	95	98	98	102
1936	103	120	87	102	99	115	92	108	97	113	97	114
1937	100	92	102	95	102	94	99	92	101	93	102	94
1938	97	91	110	105	100	95	106	101	102	97	101	95
1939	95	89	122	115	106	100	114	107	105	99	102	95
1940	92	84	136	125	105	96	115	106	107	98	103	94
1941	90	79	153	134	109	96	124	109	111	97	106	93
1942	87	69	171	135	114	90	138	109	116	91	110	87
1943	84	68	182	147	118	95	150	121	120	97	111	90
1944	80	62	201	156	121	94	162	126	124	96	114	88
1945	76	59	216	167	126	98	180	140	130	101	114	88

[1] Data used in figure 12 derived from this table.
[2] Includes only farm production share of automobile costs.

TABLE 56.—*Index numbers of total production costs and costs per unit of farm output for farm power, machinery, and labor, United States, 1910–45* [1] *(costs in current dollars)*
[1935–39 = 100]

Year	Farm-produced power		Tractors		Trucks and automobiles [2]		Other machinery		All power and machinery		All farm labor, power, and machinery	
	Total cost	Cost per unit of farm output	Total cost	Cost per unit of farm output	Total cost	Cost per unit of farm output	Total cost	Cost per unit of farm output	Total cost	Cost per unit of farm output	Total cost	Cost per unit of farm output
1910	150	197	1	1	3	4	111	146	86	113	85	112
1911	151	191	2	3	5	6	113	143	88	111	86	109
1912	183	229	4	5	7	9	117	146	102	128	95	119
1913	159	196	4	5	7	9	119	147	93	115	91	112
1914	173	208	6	7	9	11	120	145	100	120	94	113
1915	175	199	8	9	11	13	119	135	103	117	96	109
1916	174	215	11	14	16	18	121	149	105	130	100	123
1917	221	283	17	22	24	30	130	167	130	167	125	160
1918	277	330	27	32	36	46	152	181	161	192	156	186
1919	324	381	45	53	47	56	195	229	196	231	185	218
1920	332	361	64	70	60	71	208	226	208	226	206	224
1921	217	268	55	68	77	84	189	233	154	190	136	168
1922	165	185	54	61	71	88	166	187	126	142	124	139
1923	176	196	50	56	64	72	142	158	126	140	131	146
1924	182	202	53	59	68	76	141	157	130	144	133	148
1925	189	203	58	62	77	86	141	152	136	146	137	147
1926	166	175	67	71	87	94	138	145	130	137	137	144
1927	163	172	71	75	100	105	138	145	130	137	135	142
1928	152	154	76	77	102	107	139	140	128	129	133	134
1929	146	151	85	88	115	119	139	143	129	133	134	138

PROGRESS OF FARM MECHANIZATION

Year													
1930	137	144	90	95	112	118	136	143	124	131	125	132
1931	113	109	79	76	93	89	123	118	105	101	102	98
1932	80	79	69	68	83	82	107	106	85	84	78	77
1933	65	70	62	67	77	83	90	97	72	77	69	74
1934	89	113	64	81	82	104	83	105	82	104	78	99
1935	123	128	75	79	88	91	85	89	100	104	93	97
1936	90	105	84	99	96	112	90	106	90	106	92	108
1937	122	112	105	98	103	95	101	93	111	102	110	101
1938	89	84	114	109	104	99	109	104	100	95	103	98
1939	76	71	122	115	109	103	115	108	99	93	102	96
1940	82	75	130	119	108	99	115	106	102	94	104	95
1941	83	73	150	132	117	103	128	112	110	96	121	106
1942	100	79	179	141	134	106	151	119	130	102	152	120
1943	115	93	212	171	154	124	188	152	153	123	190	153
1944	138	107	252	195	175	136	234	181	183	142	226	175
1945	138	107	278	216	187	145	285	221	200	155	245	190

[1] Data used in figure 13 derived from this table.
[2] Includes only farm production share of automobile costs.

TABLE 51.—*Index numbers of total production costs and costs per unit of farm output, United States, 1910–45* [1]

[1935–39 = 100]

Year	Production costs in 1935–39 average dollars		Production costs in current dollars	
	Total	Per unit of farm output	Total	Per unit of farm output
1910	91	120	79	104
1911	94	119	81	103
1912	94	118	83	104
1913	96	119	86	106
1914	99	119	87	105
1915	99	112	91	103
1916	97	120	102	126
1917	97	124	132	169
1918	99	118	155	185
1919	102	120	176	207
1920	107	116	175	190
1921	106	131	124	153
1922	105	118	123	138
1923	104	116	132	147
1924	103	114	137	152
1925	103	111	137	147
1926	105	111	137	144
1927	105	111	138	145
1928	107	108	138	139
1929	108	111	138	142
1930	108	114	124	131
1931	107	103	97	93
1932	104	103	76	75
1933	100	108	73	78
1934	95	120	82	104
1935	97	101	91	95
1936	98	114	98	114
1937	101	93	106	98
1938	101	96	102	97
1939	103	96	103	96
1940	103	94	105	96
1941	105	92	126	111
1942	108	85	160	126
1943	110	89	193	156
1944	113	88	217	168
1945	115	89	228	177

[1] Data used in figure 14 derived from this table.

TABLE 52.—*Index numbers of price of cost factors, and of production costs and gross income per unit of output, United States, 1910–45* [1][2]

[1935–39 = 100]

Year	Cost per unit of farm output in 1935–39 average dollars	Price of cost factors	Cost per unit of farm output in current dollars	Gross income per unit of farm output in current dollars
1910	120	87	104	96
1911	119	86	103	86
1912	118	88	104	92
1913	119	90	106	94
1914	119	88	105	95
1915	112	92	103	90
1916	120	105	126	111
1917	124	136	169	172
1918	118	157	185	185
1919	120	173	207	194
1920	116	164	190	167
1921	131	117	153	117
1922	118	117	138	118
1923	116	127	147	127
1924	114	133	152	127
1925	111	133	147	138
1926	111	130	144	134
1927	111	131	145	131
1928	108	129	139	130
1929	111	128	142	134
1930	114	115	131	115
1931	103	91	93	80
1932	103	73	75	62
1933	108	73	78	73
1934	120	86	104	95
1935	101	94	95	97
1936	114	100	114	115
1937	93	105	98	101
1938	96	101	97	92
1939	96	100	96	95
1940	94	102	96	94
1941	92	120	111	115
1942	85	148	126	141
1943	89	175	156	173
1944	88	191	168	171
1945	89	199	177	178

[1] See footnote 8, page 68 and footnote 9, page 69.
[2] Data used in figure 15 derived from this table.

TABLE 53.—*Returns to all farm labor and capital, and relative returns to family and hired labor, United States, 1910–45* [1]

Year	Returns to—					Hourly returns of family workers as a percent of returns to hired workers
	Land	Working capital	Hired workers	Operator and family workers	Total	
	Million dollars	*Million dollars*	*Million dollars*	*Million dollars*	*Million dollars*	*Percent*
1910	1,577	652	757	2,489	5,475	105
1911	1,633	669	760	1,923	4,985	80
1912	1,689	630	792	2,359	5,470	95
1913	1,685	689	807	2,392	5,573	95
1914	1,749	713	805	2,475	5,742	99
1915	1,968	728	815	2,194	5,705	87
1916	2,505	728	904	2,447	6,584	90
1917	3,665	784	1,127	4,714	10,290	139
1918	3,839	989	1,335	5,641	11,804	133
1919	4,077	1,089	1,515	5,576	12,257	112
1920	2,490	1,093	1,780	5,324	10,687	92
1921	1,874	950	1,159	1,832	5,815	52
1922	2,248	732	1,122	2,831	6,933	82
1923	2,588	685	1,219	3,266	7,758	88
1924	2,992	691	1,224	2,842	7,749	76
1925	2,817	672	1,243	4,230	8,962	111
1926	2,588	706	1,326	4,071	8,691	107
1927	2,900	710	1,280	3,542	8,432	95
1928	2,765	743	1,268	4,003	8,779	108
1929	2,648	794	1,284	4,096	8,822	112
1930	1,995	798	1,134	3,004	6,931	89
1931	1,227	629	847	2,242	4,945	83
1932	866	649	584	1,390	3,489	69
1933	1,225	394	512	2,010	4,141	111
1934	1,591	375	601	2,160	4,727	105
1935	1,855	388	740	3,228	6,211	139
1936	1,995	498	880	3,204	6,577	135
1937	1,924	493	1,039	3,871	7,327	141
1938	1,697	516	1,000	3,003	6,216	115
1939	1,854	513	982	3,148	6,497	121
1940	1,904	501	1,000	3,328	6,733	126
1941	2,580	513	1,197	4,835	9,125	151
1942	3,589	656	1,566	7,404	13,215	176
1943	4,105	860	1,928	9,185	16,078	165
1944	4,355	907	2,094	8,926	16,282	132
1945	4,303	935	2,210	9,562	17,010	130

[1] Data used in figure 16 derived from this table.

TABLE 54.—*Index numbers of net land returns and value per acre, and ratio of rent to land value, United States, 1912–45*
[1935–39 = 100]

Year	Land value per acre	Net land returns per acre	Ratio of returns to value	Year	Land value per acre	Net land returns per acre	Ratio of returns to value
			Percent				*Percent*
1912	117	100	4.4	1930	138	110	5.6
1913	121	100	4.5	1931	128	66	4.6
1914	124	103	4.3	1932	107	46	3.3
1915	124	115	4.5	1933	88	64	2.8
1916	131	147	4.7	1934	92	80	3.7
1917	142	215	5.6	1935	95	95	4.5
1918	156	223	7.4	1936	99	105	5.2
1919	169	232	7.1	1937	102	103	5.5
				1938	103	93	5.4
1920	205	146	6.1	1939	101	104	4.9
1921	190	113	4.2				
1922	168	132	3.6	1940	102	107	5.5
1923	163	153	4.4	1941	103	148	5.6
1924	157	181	5.3	1942	110	209	7.2
1925	153	165	6.3	1943	120	243	9.4
1926	150	150	5.9	1944	138	264	9.5
1927	144	164	5.6	1945	152	262	9.3
1928	142	154	6.2				
1929	140	143	5.9				

TABLE 55.—*Index numbers of production costs and real returns to farm labor, United States, 1910–45* [1,2]

[1935–39 = 100]

Year	Production costs per unit of farm output in 1935–39 average dollars	Real labor returns per unit of farm output	Real labor returns per farm worker	Farm output per farm worker
1910	120	126	87	68
1911	119	100	71	72
1912	118	114	83	73
1913	119	116	85	74
1914	119	114	86	75
1915	112	94	76	80
1916	120	98	72	74
1917	124	147	107	72
1918	118	137	111	81
1919	120	115	97	83
1920	116	102	90	88
1921	131	67	52	77
1922	118	83	70	85
1923	116	91	78	87
1924	114	83	72	87
1925	111	105	93	89
1926	111	102	92	90
1927	111	93	85	92
1928	108	97	93	96
1929	111	102	96	94
1930	114	84	79	93
1931	103	67	69	102
1932	103	52	53	100
9933	108	73	67	92
1934	120	84	67	80
1935	101	97	92	94
1936	114	114	96	84
1937	93	103	112	108
1938	96	91	97	106
1939	96	95	103	108
1940	94	96	108	112
1941	92	118	141	120
1942	85	134	179	134
1943	89	154	203	132
1944	88	140	196	140
1945	89	144	206	143

[1] See footnote 11, page 73.
[2] Data used in figure 17 derived from this table.

TABLE 56.—*Real income per farm worker and per industrial worker, United States, 1910–45* [1] [2]

Year	Real income per farm worker	Real income per industrial worker	Real income per industrial worker adjusted for unemployment	Index numbers (1935–39=100)		
				Real income per farm worker	Real income per industrial worker	Real income per industrial worker adjusted for unemployment
	Dollars	*Dollars*	*Dollars*			
1910...	334	868	855	87	76	89
1911...	275	826	792	71	72	83
1912...	319	833	813	83	72	·85
1913...	328	845	823	85	74	86
1914...	333	838	791	86	73	83
1915...	292	864	813	76	75	85
1916...	276	890	886	72	77	93
1917...	413	889	889	107	77	93
1918...	427	994	994	111	87	104
1919...	373	958	958	97	83	100
1920...	347	987	974	90	86	102
1921...	200	964	856	52	84	90
1922...	272	985	918	70	86	96
1923...	303	1,044	1,026	78	91	107
1924...	277	1,043	995	72	91	104
1925...	359	1,034	1,015	93	90	106
1926...	355	1,046	1,036	92	91	108
1927...	330	1,057	1,020	85	92	107
1928...	359	1,076	1,034	93	94	108
1929...	369	1,093	1,060	96	95	111
1930...	304	1,050	956	79	91	100
1931...	266	1,037	871	69	90	91
1932...	203	948	720	53	83	75
1933...	260	978	734	67	85	77
1934...	257	1,024	799	67	89	84
1935...	356	1,080	864	92	94	90
1936...	370	1,141	958	96	99	100
1937...	433	1,183	1,017	112	103	106
1938...	375	1,123	921	97	98	97
1939...	396	1,217	1,022	103	106	107
1940...	417	1,273	1,095	108	111	115
1941...	544	1,424	1,310	141	124	137
1942...	690	1,593	1,593	179	139	167
1943...	785	1,755	1,755	203	153	184
1944...	757	1,859	1,859	196	162	194
1945...	797	1,758	1,758	206	153	184

[1] Yearly returns to labor per farm worker were deflated by the index of prices paid by farmers for commodities used in living (1935–39 = 100), and average annual wages per industrial worker were deflated by the B.L.S. Cost-of-living index (1935–39 = 100). Adjustment of industrial workers' income for unemployment was made by multiplying the percentage employed in labor force by the unadjusted income per worker.

[2] Data used in figure 18 derived from this table.

...CHANGES IN
American Farming

By SHERMAN E. JOHNSON • Bureau of Agricultural Economics

MISCELLANEOUS PUBLICATION No. 707

UNITED STATES DEPARTMENT OF AGRICULTURE
WASHINGTON, D. C. DECEMBER 1949

PREFACE

This study was undertaken in the belief that a first requisite for intelligent action on our peacetime production problems is an understanding of the nature and strength of the forces that have shaped the course of agricultural production in recent years. These forces will continue to influence production in the years to come. But new forces will be injected. Some of these are already on the horizon—the mechanical cotton picker is an example. Others cannot be foreseen. But constant change must be expected; and agriculture must adapt itself to changes that are inevitable and, for the most part, desirable.

The revolutionary changes in farm production since World War I are not reversible. Policies developed for the present and for the prospective production situation, therefore, should reflect recent developments and those that can now be foreseen. They cannot be fashioned from the pattern that prevailed before the age of mechanical power and before other fundamental changes had taken place.

The background data used for this analysis are largely the statistics contained in the crop and livestock reports and other production releases of the Bureau of Agricultural Economics.

Many workers participated in this study. Several reports were issued that deal with changes in specific commodities, or with other phases of changes in farming. Basic to this summary report is the special study reported in Farm Production in War and Peace, by Glen T. Barton and Martin R. Cooper, issued by this Bureau. Others who worked on specific parts of the study are: Neil W. Johnson, C. W. Crickman, Carl P. Heisig, E. L. Langsford, Donald B. Ibach, Olav Anderson, R. D. Jennings, S. W. Mendum, E. G. Strand, Weber Peterson, W. D. Goodsell, R. W. Jones, J. R. Ferrell, E. R. Ahrends, L. Jay Atkinson, R. W. Hecht, A. P. Brodell, K. L. Bachman, and Della E. Merrick who assembled and helped to prepare materials for the entire project. Paul L. Koenig, R. K. Smith, and the late J. B. Shepard were especially helpful in appraising background data.

Changes in American Farming

By SHERMAN E. JOHNSON, *Assistant Chief, Bureau of Agricultural Economics*

CONTENTS

	Page
Significance of recent changes:	
Record farm output	1
The war years	3
World War II increases compared with World War I	10
The early postwar years	11
Foundation for increased production:	
Significance of interwar years	12
Changes in the use of farm machines	13
Changes in land use, conservation, and fertilizer practices	20
Changes in crops	27
Foundation for increased production—Continued	Page
Changes in livestock production	45
Changes in farm sizes and ownership	51
Incentives for increased production	56
Implications of recent and prospective changes:	
Prospective changes	58
Effect of prospective changes on farm output	62
Output in relation to potential markets	63
Adjusting market outlets to balanced farm output	66
Profitable and abundant farm production	67

SIGNIFICANCE OF RECENT CHANGES

RECORD FARM OUTPUT

Farmers in the United States achieved a remarkable production record in the war years. The increase in total farm output from 1935–39 to 1944 was twice as large as during the entire period from 1919–23 to 1935–39. This was accomplished without significant expansion in the acreage of cropland, and despite scarce supplies of labor, machinery, and farm materials. The high level reached during the war has been maintained in the early postwar years, and the 1948 output was an all-time record.

The output of farm products available for human use for the three full war years 1942–44 averaged 128 percent of the prewar years 1935–39. In 1946 the level was 133 percent of prewar. Despite a short corn crop the output in 1947 averaged 129 percent of prewar. If the corn crop had been as large as might have been expected with average weather the total output in 1947 would have been about 132 percent of prewar. With a bumper corn crop in 1948 farm output reached the record total of 140 percent of 1935–39 (fig. 1, p. 2).[1]

This increase in output constitutes an unprecedented break from previous trends. Usually, changes in farming develop very slowly. They are often unnoticed until the record over a period of years is examined. Even such major innovations as the tractor and complementary machines adapted for mechanical power were introduced so gradually that they escaped special attention until their cumulative effects became unusually pronounced.

Sometimes extremely favorable or unfavorable weather brings large year-to-year changes in production. For example, 1934 was a year of catastrophic drought, and farm production was much lower than in the preceding years. In 1942, growing conditions were unusually favorable. In fact, consistently good weather was experienced in the three war years 1942–44 compared with the average of the years 1935–39. But those prewar years reflect weather conditions that were less favorable to farm production than is the expectancy over a period of years. And although weather factors were more favorable in 1942–44 than longer-time expectancy, other forces were responsible for most of the increase in production.

Considering the average of the years 1942–44, it appears that about one-fourth of the total increase in production can be accounted for by weather conditions that were more favorable than in the prewar years 1935–39. This means that with normal weather farm output in 1942–44 would have averaged about 120 percent of 1935–39.

Obviously then, only a rather small part of the wartime increase in production can be explained by favorable weather. And it follows that average weather alone would not bring a return to prewar production levels. Extremely unfavorable weath-

---

[1] For a detailed explanation of how this index measure of physical production is constructed see GLEN T. BARTON and MARTIN R. COOPER, "FARM PRODUCTION IN WAR AND PEACE," U. S. Bur. Agr. Econ. F. M. 53, 85 pp., illus. 1945. [Processed.] (Supply exhausted.)

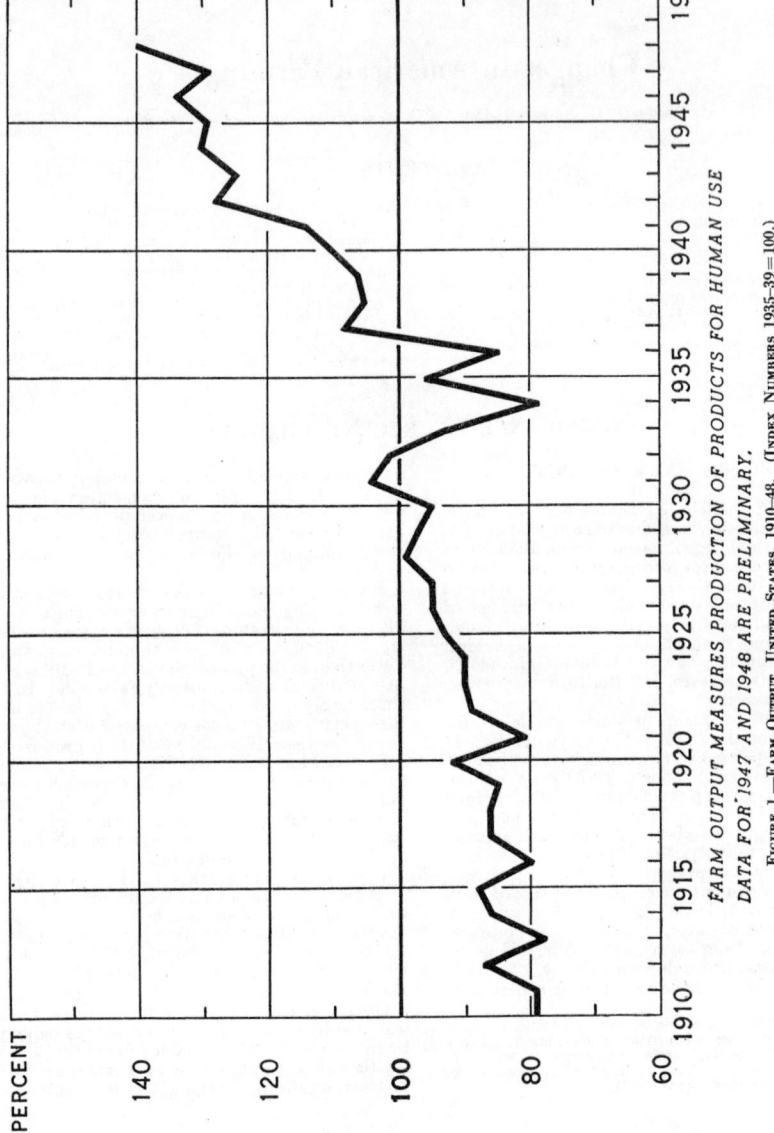

FARM OUTPUT MEASURES PRODUCTION OF PRODUCTS FOR HUMAN USE
DATA FOR 1947 AND 1948 ARE PRELIMINARY.

FIGURE 1.—FARM OUTPUT, UNITED STATES, 1910–48. (INDEX NUMBERS 1935–39 = 100.)

Farm output in the war years 1942–44 was 28 percent higher than in the prewar years 1935–39. The output in 1948 was 40 percent above prewar.

er would reduce output temporarily. But agriculture experienced a production revolution during the war years, and a large part of the change is irreversible. It will persist under peacetime conditions. To understand what has taken place it is necessary to analyze what happened in the war and early postwar years, to compare this experience with the record in World War I, and to trace the foundation for production increases that was laid in the interwar years.

THE WAR YEARS

COMBINATION OF FORCES BACK OF HIGHER OUTPUT

War and wartime needs for food, and the doubling of the prices received by farmers for their products, furnished the driving force for increased production, but a combination of favorable physical circumstances made the large-scale increase possible. Potential production capacity had been built up over the several years previous to the outbreak of World War II. This increased capacity had its origin in several factors, and each contributed to the higher wartime output. By a fortunate conjuncture of circumstances widespread progress in mechanization, greater use of lime and fertilizer, cover crops, and other conservation practices, use of improved varieties, a better balanced feeding of livestock, and more effective control of insects and disease, had all gathered momentum over the several years preceding World War II. Their current effects were obscured by the drought and depression of the 1930's, but developments had reached a stage where these improvements could be effectively combined and used in an all-out production effort. The result was an unprecedented increase in output.

The joint effects of these technological improvements on the volume of production may be illustrated by comparing them with the effects of the flow of water in its several tributaries on the water level of a large river. If water rises to flood stage in one of the tributaries this will, of course, increase the water level in the main stream, but if the tributary is small the effect may be scarcely noticeable when its flood reaches the main channel. Similarly, the effect of single improvements in farm production, that are important by themselves, are scarcely perceptible in their effect on total production. But if all the tributaries of a large river reach flood stage at the same time, the water in the main channel also rises to flood stage, and the change in the water level does not escape notice. In a sense this is the effect that adoption of the accumulation of technological improvements had on farm production in the years of World War II.

But one might make the comparison somewhat differently, and more correctly, by saying that the production-increasing potentialities of improvements that were made over a decade, and that normally would have been diverted gradually into the production stream, were held back by the drought and depression of the 1930's. It was the breaking of these restraints that caused the flood of production in the war years—in the same way that a simultaneous breaking of dams on several tributaries will cause a river to reach flood stage from water that was accumulated from a normal flow at the source.

ACCUMULATION OF POTENTIAL CAPACITY

As this accumulation of potential production capacity had escaped notice, the increase that was achieved was much larger than could have been forecast from past trends. It was much greater than the expansion that took place in World War I, because there was no similar accumulation of potential improvements at that time. Perhaps none familiar with the South would have been so rash as to forecast, in the fall of 1941, that the acreage of peanuts picked and threshed in 1942 would be 177 percent of 1941, and that production for the years 1942–44 would average 175 percent of 1935–39. Likewise, none from the Corn Belt would have dared to forecast in 1941 that the production of soybeans harvested for beans in the years 1942–44 would be 338 percent of the production in 1935–39. Figure 2 shows the average 1942–44 production compared with prewar for some products in which major changes occurred. An optimistic advance estimate of wartime production probably would have averaged less than half of the increase that actually was achieved.

Mechanization was one of the most influential factors back of the increased output of farm products. The number of tractors on farms had gradually increased from less than 250,000 in 1920 to nearly 2,500,000 in 1945. Use of mechanical power and complementary equipment usually means more total production, but its most important effect is that a much larger share of the product goes to market. As mechanical power is substituted for draft animals, the land formerly used for horse and mule feed becomes available for producing commodities for human use. The shift to mechanical power from 1918 to 1945 made available about 55,000,000 crop acres, or about 15 percent of the available cropland, for the production of marketable commodities. In World War I this large area of cropland and millions of acres of pasture had to be used for producing feed for horses and mules.

Greater use of fertilizer and lime was another influential factor in stepping up farm output.

Measured in plant nutrients (N, P_2O_5, K_2O), the total consumption of commercial fertilizer in 1945 was 95 percent above the quantity used in the prewar years, 1935–39. Application of liming materials was more than three times as large as in the prewar years. Based on estimates of additional output from increased use, it appears that the increased production resulting from the additional use of lime and fertilizer in 1945 accounted for about 15 percent of the total increase in output since 1935–39.

Crop improvements were another notable source of increased output. Use of hybrid seed increases the yield expectancy of corn about 20 percent.

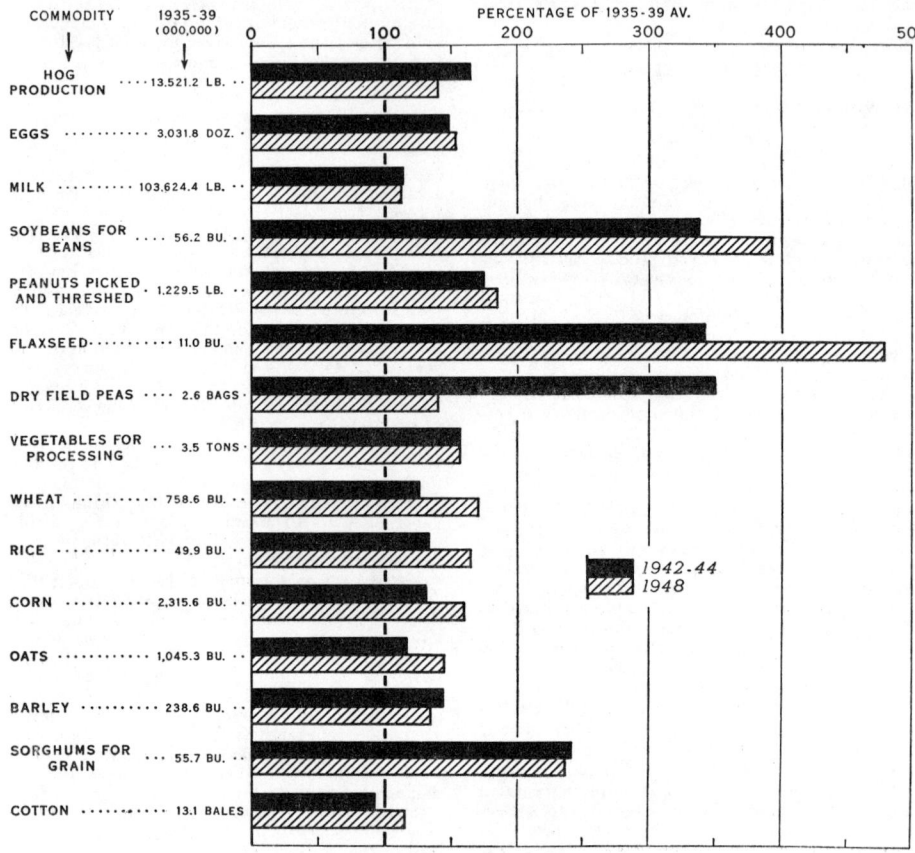

FIGURE 2.— PRODUCTION OF SELECTED AGRICULTURAL COMMODITIES IN THE UNITED STATES. PREWAR (1935–39 AVERAGE) AND WARTIME (1942–44 AVERAGE) AND 1948.

Production of nearly all the principal agricultural commodities increased in the war years. The sharp advances in oil crops reflect responses to the acute need for fats and oils. The phenomenal increase indicated for dry field peas is of less over-all importance than some of the smaller percentage increases registered from a larger prewar base such as those for eggs and milk. The expansion of wheat production has occurred mostly in the postwar years.

New varieties of oats adapted for use in the Corn Belt and Lake States, and rust-resistant improved winter varieties for the Southern States, were widely adopted in the later war years. Continued progress was made in shifting to the higher yielding and more nutritious legume hays.

Total land used for intertilled crops increased by about 6,000,000 acres, or 4 percent, from 1935–39 to 1944. Meanwhile, the total cropland used for crops increased about 9,000,000 acres, or 3 percent. This is a rather small change and is therefore a minor factor in increased output.

The building up of both livestock numbers and feed supplies in the years immediately preceding the war, and in 1942, made possible larger marketings of livestock and livestock products, especially in 1943 and 1944. Livestock production was increased also by the feeding of accumulated supplies of wheat, and of some imported wheat and feed grains. But feeding of both the accumulated supplies and the larger imports accounted for only about 10 percent of the total concentrates fed to livestock in the year ended October 1, 1944. Thus by far the largest proportion of the livestock output came from current production of grain, forage, and pasture; and from an increase of more than one-half in oilseed meals for livestock feed.

Fortunately there were no major outbreaks of either plant or animal diseases or of insect damage in the years of World War II. That no livestock diseases reached epidemic proportions despite record-breaking inventories of livestock was not only good fortune; it was an indication of the effectiveness of modern control methods, and of the vigilance of both farmers and technicians in controlling sporadic outbreaks. Insect damage was held to low levels despite shortages of such important insecticides as rotenone and pyrethrum.

With financial and patriotic incentives as encouragement, and with education and persuasion centered on virtually all-out production of strategic products, farmers and their families worked long hours and often utilized, to the best of their knowledge, every possible means of increasing output. Their efforts bore fruit so well that, even though about one-fourth of the food output went to military and other war-emergency uses, there was food enough in 1944 and 1945 to provide our civilians with a per capita food consumption 12 to 14 percent higher than took place in the prewar years, 1935–39. In somewhat different terms, the output of food in 1944 was enough to feed about 50,000,000 more people than were fed by the average quantity produced in 1935–39, assuming the same dietary levels for both periods.

UNFAVORABLE FACTORS MINIMIZED

That more people could be fed resulted partly from a change in the production pattern—more oil crops, beans, and peas, and less cotton—and from a more complete utilization of the output. But the shifts in production that were necessitated by war needs, made increases in the total volume of production more difficult (fig. 2, p. 4). Production per acre or per animal is usually lowered when a product is grown on land that is less suited for its production, or by growers who have insufficient experience.

The wartime increases in production were achieved with a constantly shrinking labor supply. The total farm population dropped from 30,000,000 in 1940 to 25,000,000 in 1945. In many small farm areas the decrease in farm population represented a correction of under-employment on farms that existed before the war; but in most of the commercial farming areas the result was a labor shortage.

Figure 3, page 6, shows the downward trend in farm employment and the contrasting sharp upward trend in production per worker. In 1944 farmers had 8 percent fewer workers than in 1935–39. Many of the hired workers who were available did not have the strength and skill that are usually considered necessary for farm work. But farmers and their families worked longer hours and, somehow the job was done.

HIGHER PRODUCTION PER UNIT OF RESOURCES

Figure 4, page 7, shows the changes in production per acre in relation to the changes in acreage of total cropland. The effects of all the forces that have resulted in higher production per acre are combined in this index. Similarly, the line showing production per animal unit of breeding livestock, in figure 5, page 8, combines into one summary figure all the forces that have increased livestock production per animal. It shows the trend in production per unit of breeding stock. The shift to mechanical power and the increase in production per acre have made it possible also to increase the number of animal units of productive livestock. Thus the total increase in livestock production is derived both from larger numbers of breeding stock and from the higher output per unit of breeding stock as shown in figure 5. On the other hand, figure 4 shows that increased crop production is largely the result of higher production per acre because changes in the acreage of cropland have been relatively small.

The higher output per acre and per animal, combined with mechanization, made possible the larger output per worker. The three series—production per acre, per animal, and per worker—summarize the startling changes that have taken place in farming since the years of World War I.

During the time that progress in technology has increased output per man so greatly in agriculture, the same phenomenon has occurred in industry.

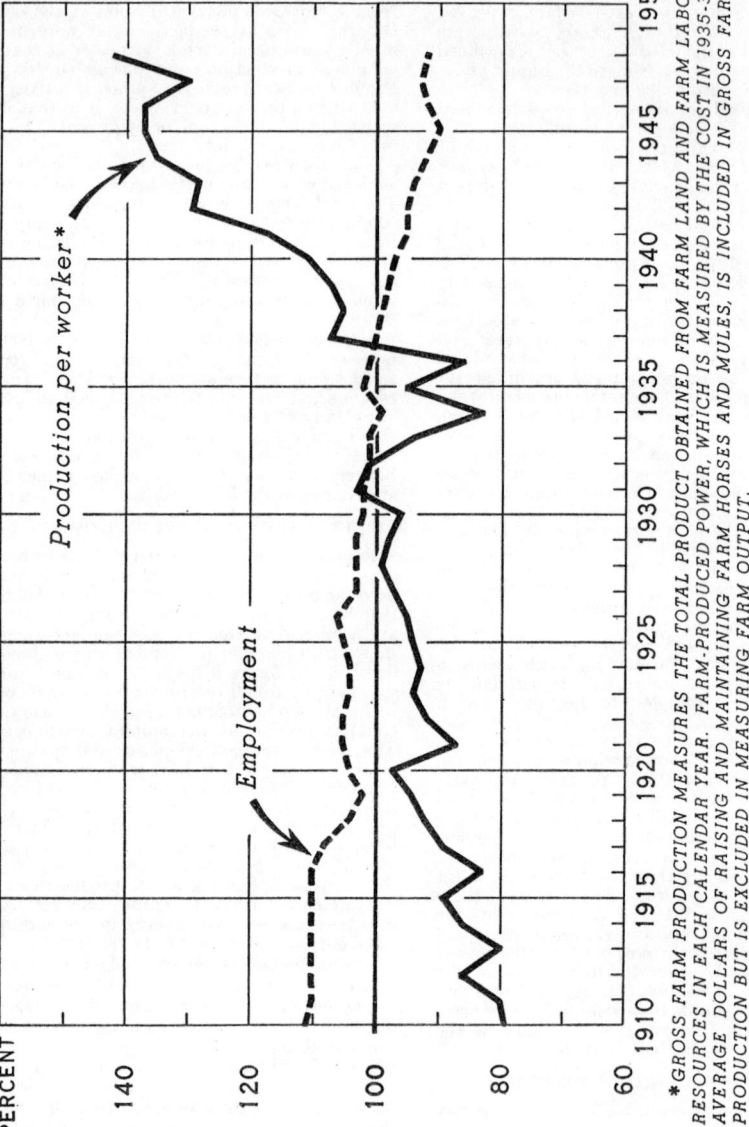

FIGURE 3.—FARM EMPLOYMENT, AND GROSS FARM PRODUCTION PER WORKER, UNITED STATES, 1910–48. (INDEX NUMBERS 1935–39=100.)

*GROSS FARM PRODUCTION MEASURES THE TOTAL PRODUCT OBTAINED FROM FARM LAND AND FARM LABOR RESOURCES IN EACH CALENDAR YEAR. FARM-PRODUCED POWER, WHICH IS MEASURED BY THE COST IN 1935-39 AVERAGE DOLLARS OF RAISING AND MAINTAINING FARM HORSES AND MULES, IS INCLUDED IN GROSS FARM PRODUCTION BUT IS EXCLUDED IN MEASURING FARM OUTPUT.

DATA FOR 1947 AND 1948 ARE PRELIMINARY.

Acceleration during the war of the long-time downward trend in farm employment, coupled with marked wartime increases in farm production, resulted in record levels of production per farm worker.

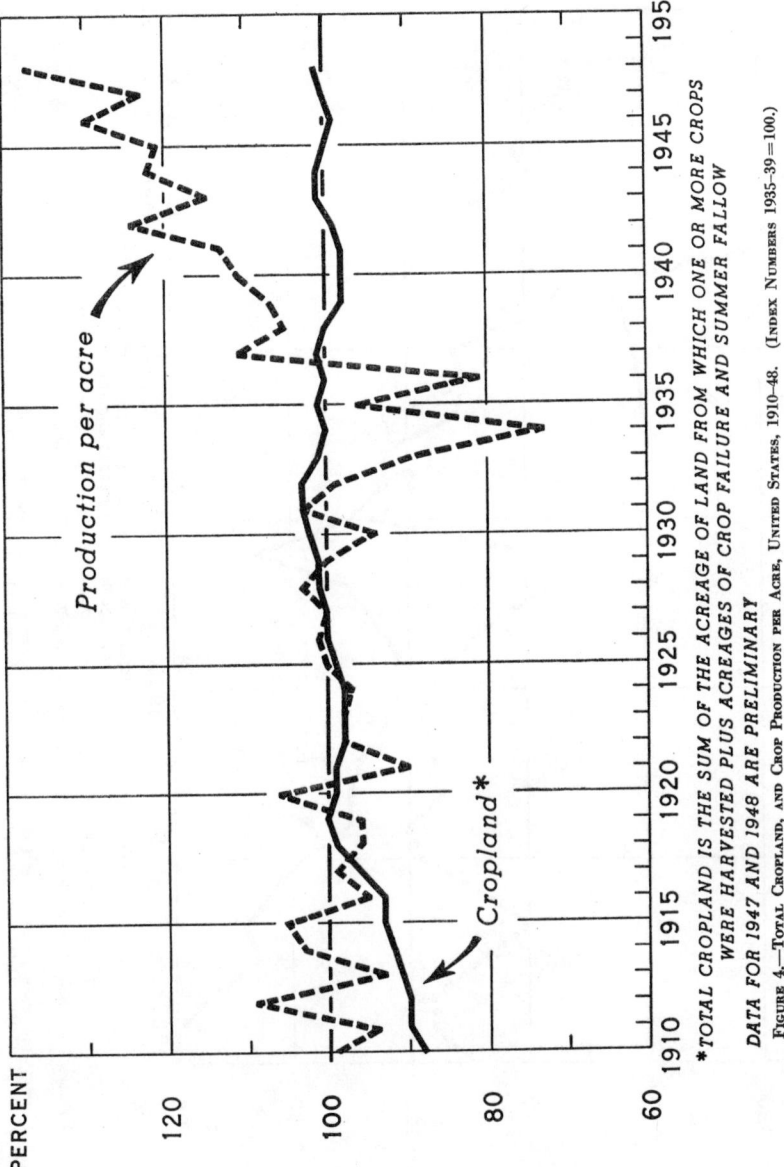

FIGURE 4.—TOTAL CROPLAND, AND CROP PRODUCTION PER ACRE, UNITED STATES, 1910–48. (INDEX NUMBERS 1935–39=100.)

Increased crop production per acre was by far the most important single factor responsible for record wartime farm production. Total cropland has changed very little during the entire period.

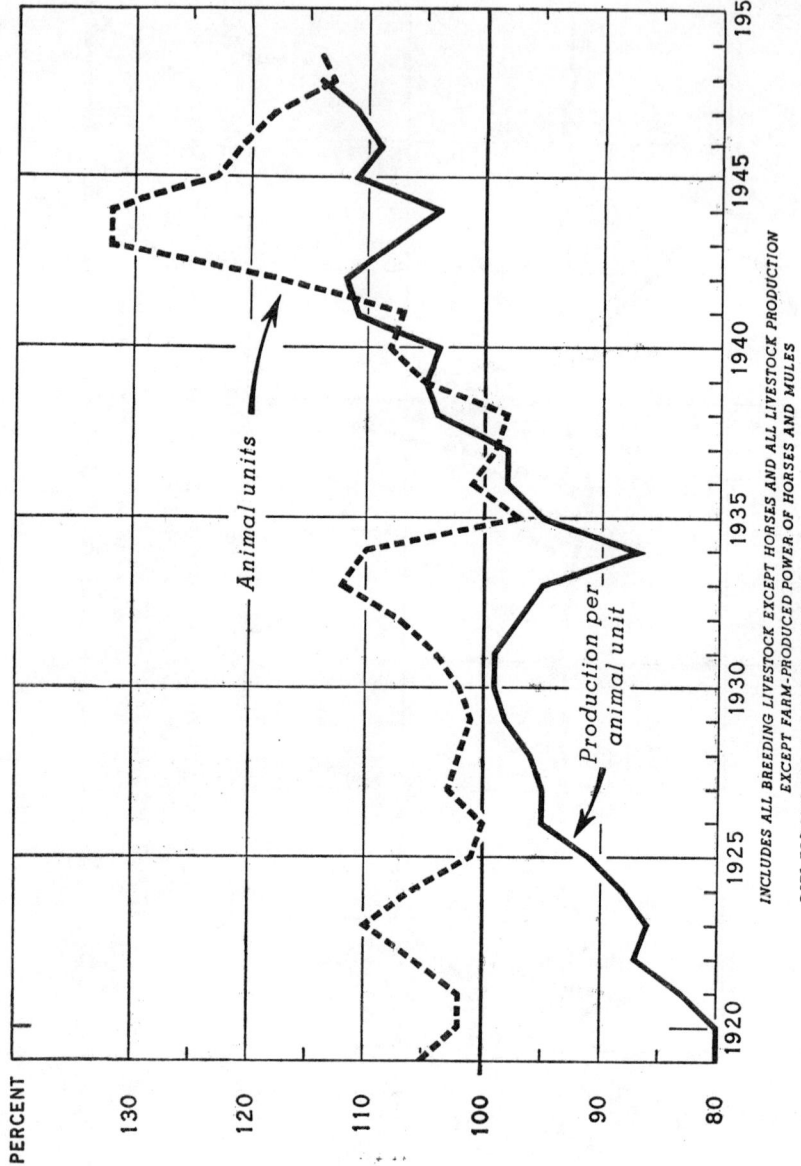

FIGURE 5.—ANIMAL UNITS OF BREEDING LIVESTOCK AND LIVESTOCK PRODUCTION PER BREEDING UNIT, 1919–48. (INDEX NUMBERS 1935–39=100.)

Livestock production per animal unit of breeding stock has shown an upward trend since 1920, and the animal units of breeding livestock also have increased. The larger volume of livestock production therefore has resulted both from increases in number of breeding animal units and in production per unit.

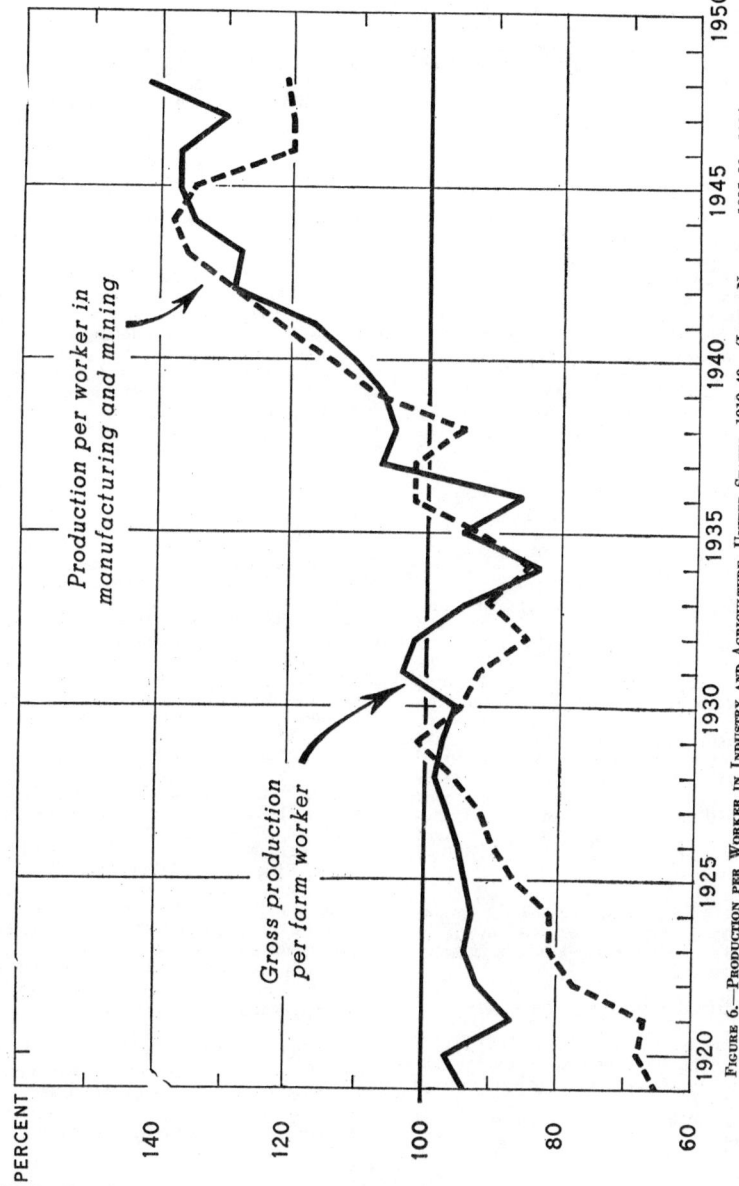

FIGURE 6.—PRODUCTION PER WORKER IN INDUSTRY AND AGRICULTURE, UNITED STATES, 1919-48. (INDEX NUMBERS 1935-39=100.)

Increases in production per worker in agriculture matched those of industrial workers during the war, and have remained at a relatively higher level; but production per worker in industry has risen more rapidly than that of farm workers over the last quarter-century.

Figure 6, page 9, shows a somewhat more rapid increase in production per worker in the manufacturing and mining industries than in agriculture over the last quarter-century. But if data were available for the service industries they probably would show a less rapid climb. Increases have occurred in all sectors, however.

This poses a key question for the postwar years. Will peacetime industry and service occupations expand sufficiently to absorb at satisfactory wage levels (1) the workers currently displaced by technological progress in both agriculture and industry and (2) the net supply of new workers (after allowing replacements for death and retirement) that enter the labor market each year? This is a crucial question. But before it is discussed in relation to peacetime agricultural production it is necessary to examine our wartime and early postwar records more closely. And first of all, it seems desirable to compare changes that took place in World War II with those of World War I, because production responses were so different despite the similarity of incentives to increase farm output.

WORLD WAR II INCREASES COMPARED WITH WORLD WAR I

WORLD WAR I FOLLOWED PERIOD OF EXPANDING MARKETS

World War I followed a period of expanding agriculture, but the markets for farm products had caught up with the expanded production before that war started in Europe in 1914. The export market had been large and profitable over a period of years, and the increasing population in this country had gradually absorbed a larger proportion of the total farm output. Machinery suitable for use with animal power was fairly well stabilized. Tractors were only in the early stages of adaptation for farm use. Most of the good virgin lands had been occupied before the beginning of the war. Criticisms of the high cost of living with special reference to food and fiber were being heard in the cities. A back-to-the-land movement was advocated by many people. Some of the current literature emphasized the importance of increasing output per acre in view of the fact that little new land was available for settlement and exploitation.[2]

INCENTIVES FOR HIGHER OUTPUT

Prices and income incentives in World War I were favorable to the expansion of output. Taking the years 1910–14 as a base, prices received by farmers for farm products in 1918 were 204 percent of those prewar years. By 1919 they were 215 percent. Prices of products bought by farmers also increased during the war, but at slower rates; and in 1918 and 1919 net incomes to farm operators reached new high levels. But with financial and patriotic incentives similar to those of World War II, and with the need for food just as urgent, the volume of output for human use increased only about 5 percent from the prewar years 1910–14 to 1918–19.

What are some of the reasons for the differences in the increases of production in the two war periods?

Although the most productive virgin lands had been brought into cultivation by the beginning of World War I, some new land was available for exploitation. Harvested cropland increased more than 32,000,000 acres from the years 1910–14 to the years 1918–19. This is more than twice as large an increase in harvested cropland as occurred from the prewar years 1935–39 to the war years 1942–44. It should have resulted in greater output at that time if other factors had been favorable.

The foods receiving most emphasis in World War I were wheat and meat. In World War II there was much less emphasis on wheat until after the end of hostilities, and relatively more on oil seeds, beans, and peas; and on meat, eggs, and dairy products. Many of the products emphasized in World War II were new products on many farms, so increases were somewhat more difficult to achieve than expansion of the products that are normally produced.

In World War I the wheat acreage was expanded to new lands in the drier wheat-producing areas, and it was substituted for other crops on the older lands of the humid areas in the Corn Belt and the Eastern States. The acreage of wheat harvested in the five major Corn Belt States in the years 1917–19 was 129 percent of the 1910–14 figure. A new peak of harvested wheat acreage in the United States was reached in 1919, with 73.7 million acres.

Cattle numbers were at the low point of a cycle before the beginning of World War I in 1914, and steps were taken to encourage expansion of cattle production, both on eastern farms and on the ranges of the Western States. The Forest Service was requested to increase the numbers allowed under grazing permits in the national forests. Under the stimulus of high prices and other incentives, cattle numbers on January 1 increased from a low of 56,000,000 in 1912 to a high point of 73,000,000 in 1918. For World War II the high point in cattle numbers was reached January 1, 1945, with 86,000,000 on farms and ranches. But in comparing the two periods one must recall that the decrease in horse and mule numbers had re-

[2] See SPILLMAN, W. J. SOIL CONSERVATION, U. S. Dept. Agr. Farmers' Bul. 406, 15 pp. 1910.

leased land equivalent to the acreage needed for maintaining 16,500,000 head of cattle and calves.

Hog production was stimulated in World War I, partly by a promise to maintain prices of hogs at a fixed ratio to corn, at Chicago. In 1919 the number of all hogs on farms January 1 had increased 21 percent—from 53,000,000 in 1914 (a low point in the hog cycle) to 64,000,000. In World War II, numbers on January 1 increased 68 percent from 50,000,000 in 1939 to a high point of 84,000,000 in 1944.

Despite the emphasis on wheat and meat in World War I, and the substitution of these products for other products in the farm economy, wheatless and meatless days were proclaimed. Adequacy of basic food supplies remained a critical war problem.

FORCES THAT RETARDED EXPANSION

Weather was not so favorable during the years of World War I as in World War II. The wheat crop of 1916 was reduced by adverse weather and damage by black stem-rust. In 1917 the crop was even smaller than in 1916 because of severe winter killing of winter wheat and of drought in some spring-wheat areas. Carry-over reserves were badly depleted by exports to Allied Nations. The wheat crops were nearly a third larger in 1918 and 1919 than in 1917, but these gains in production became available chiefly after the end of the war.

Although manufacture of farm machinery was not limited by Government order as in World War II—in fact, its manufacture was encouraged until the fall of 1918—mechanical power was still of minor importance. Tractor power contributed only 2 percent of the combined work-animal and tractor-power units on farms, in 1919. Farming principally with animal power meant that much land that now produces marketable products was needed for horse and mule feed, and that reserve power for long hours of field work, and for timeliness of operations, was not available.

Because potash fertilizers had been imported from Germany, there was a serious shortage of this plant nutrient during World War I. The available supply in 1918 was only about one-fourth of that used annually in the immediate prewar years. Nitrogen supplies were also short. In fact, total use of fertilizer was considerably reduced in the early part of that war and, though it increased somewhat in 1918–19, it did not reach the levels of 1914 at any time during those war years.

Plant and animal diseases and insect pests took a proportionately greater toll of potential production in World War I than in World War II. The boll weevil caused extensive damage to cotton in all of the years of the first war. Black-stem rust seriously affected wheat production in 1916. Outbreaks of hog cholera were numerous from 1911 to 1915 and again in 1919.

The situation in World War I as compared with World War II can be summarized in this way. More sod land that could be broken up and put to use was available; but on the intensive side (increases in output per acre and per animal) production was limited by shortage of fertilizer, lack of mechanical power, damage by plant and animal pest and disease, and somewhat unfavorable weather. There were no innovations such as hybrid corn that could be seized upon to increase production quickly. The background of agricultural research was more limited, and organized extension teaching was in its early stages in most States. In other words, the same favorable conjuncture of circumstances for agriculture that prevailed in World War II did not develop in World War I.

It should be emphasized that, aside from weather, all of the physical conditions more favorable to increased production in World War II have grown out of invention and research, and have been spread by the education and operations programs of the interwar and World War II years. The cumulative value of agricultural research and of extension teaching in making possible the increase in farm output in World War II can hardly be overestimated. And the stimulation given by operations programs to conservation practices and increased use of lime and fertilizer must not be overlooked.

THE EARLY POSTWAR YEARS

IMPACTS OF FOOD RELIEF PROGRAMS

The record now includes the first three production years following the end of hostilities. During the first two of these years the pressure on supplies of food grain intensified, and wheat again took first place in meeting the food needs of hungry people. In 1948 both Europe and North America had favorable growing weather. And farmers in Western Europe produced a much larger crop than in 1947. This resulted in a lessening of pressure on available food supplies and in accumulation of stocks of food and feed grains in this country.

In view of the urgency regarding food relief at that time it was fortunate indeed that growing conditions in this country were exceptionally favorable for both food and feed grains in 1946. Wheat and corn production made new records. In 1947 the wheat crop exceeded even the high record established in 1946; the total outturn was 1.4 billion bushels. But the corn crop was nearly a billion bushels smaller than in 1946. The relative shortage of feed grains severely limited livestock production for 1948. But the 1948 corn crop was an all-time record of 3.6 billion bushels, and a wheat crop of 1.3 billion bushels was the second largest on record.

Farm prices moved upward rapidly in 1947. In January 1948 they were nearly three times as high as in the prewar years 1935–39. After some decline in February and March they recovered nearly to January levels in July 1948. But another drop began in the summer of 1948 and continued without interruption to March 1949. For the year 1948 farm prices averaged 268 percent of 1935–39, which was 42 percent above the level in 1945.

Prices paid for goods and services used in farm production also climbed rapidly in 1946 and 1947. For the year 1948 they averaged 201 percent of 1935–39. This was 47 percent higher than in 1945. There was a slight decline in the summer of 1948, but this was largely the result of lower feed prices. Other production expenses remained at higher levels.

The margin between costs and selling prices was favorable to most producers in 1946 and 1947 and net incomes to farm operators reached successive peak levels in those years. In 1948 the margin became less favorable and net income to farm operators went down for the first time in 10 years. Although the 1948 net income was a little over 2 percent less than in 1947, it was still the second highest on record. It is apparent that income incentives to high-level production were good in the three postwar years 1946–48, and more farm labor was available than in 1945. Supplies of farm machinery and commercial fertilizer were larger than in previous years, although supplies were not sufficient to meet all demands.

POTENTIAL PRODUCTION CAPACITY

Supplies of production goods are now much more plentiful than they were during the war and early postwar years. If income incentives remain definitely favorable, are we likely to see a further increase in output in the next 3 to 5 years? Or are we now on a high plateau with greater prospects of a downward slide than of a further rise? These are vital questions in relation to market outlets for certain farm products. And they become extremely significant if we do not make progress in achieving a stable peace.

The corn crop of 1947 indicates the potential effects of weather on farm output. If we should have a drought comparable to that of 1934 the output might drop as much as 20 percent in any one year. Moreover, such a disaster would necessitate the selling of breeding livestock which would affect output in later years.

Given average weather and sufficient time to make the changes that would be necessary in the farm-production plant, the only practical limit to our farm output would be the labor and capital that we consider profitable to use in agricultural production. In peacetime this depends mainly upon decisions of individual farmers, which in turn are based upon their confidence in continued favorable market outlets for farm products.

Supply problems are emerging in some farm products. The transition toward peacetime production in the next 4 to 5 years seems likely to require changes in the direction of raising less wheat, corn, and soybeans, and more emphasis on hay and pasture, in the interest of conservation and of producing more meat and milk.

FOUNDATION FOR INCREASED PRODUCTION

SIGNIFICANCE OF INTERWAR YEARS

An examination of production changes in the interwar years clearly indicates that the foundation for the wartime and early postwar increases in production was built up in the 20-year period between the two World Wars. It was built on the tremendous improvements that have been made in mechanization, in land use, in conservation, fertility, and cultural practices, in development of new crops and new varieties, and in increasing the efficiency of livestock production.

The cumulative effect on the total volume of production was not realized because of drought and because markets were not available for larger quantities of farm products. We have seen that although farm output increased slowly during the 1920's, it dropped precipitously during the drought of the early 1930's (fig. 1, p. 2). Later, production reached its new peak in 1937 and remained relatively high in 1938 and 1939.

Except for the severe drought years, farmers obviously could have produced more in the 1930's if markets had been available at profitable prices. But despite the brake placed on increases in production by the low prices and resulting crop-adjustment programs, it seems evident that the total volume would have risen sharply in the early 1940's even without the incentive of wartime prices and needs. The production potential was built up in the late 1930's to the point where accelerated increases in production were almost inevitable.

The stimulation of war needs—the patriotic urge to increase production, the doubling of farm prices, and the tripling of net incomes to farmers—pushed production faster and farther than it would have gone under peacetime conditions. But this greater acceleration was physically possible only because of the foundation that had been laid in the interwar years.

CHANGES IN THE USE OF FARM MACHINES

EFFECTS OF SHIFTS TO MECHANICAL POWER

The rapid shift from animal power to mechanical power for farm production in the interwar period constituted one of the most important changes that has ever taken place in American agriculture. It was a cornerstone in the foundation for increased production.

One result of this change was a transfer of resources from the production of power on the farm to the production of livestock and crops for sale in the market. This transfer released about 55,000,000 acres of cropland for the production of marketable farm products. That, of course, is the most startling effect that mechanization has had on American farms.

But other effects are also striking. The shift to mechanical power and complementary equipment brings an increased output per worker both by enabling him to do the job faster and by doing a better job of tillage or other operation, and in that way realizing more benefit from other improved practices. The physical burden of farm work is lessened because the drudgery of hand labor is eliminated. Fewer workers are needed in farm production. Physical strength becomes relatively less necessary, and mechanical skill becomes more necessary, in the performance of farm work.

CONTRIBUTION TO TIMELINESS IN FARM OPERATIONS

The effects of the greater timeliness in farm operations that is possible by mechanization are difficult to measure, but obviously the result is both greater total production and higher quality of farm products. As power equipment can cover more acres per hour and can be used longer hours if necessary, it has enabled farmers to do critical jobs without the delay that frequently occurred when horse equipment was used. For example, the quick coverage possible in spraying potatoes with a multiple-row power sprayer may salvage a crop that otherwise would be seriously damaged.

Perhaps the best illustration of the contribution of mechanical power to timeliness of operations is the experience in the Corn Belt with unfavorable weather at planting time in each of the years 1943 to 1945 and again in 1947. The spring of 1943 was exceptionally wet, so planting was seriously delayed in many areas. Illinois usually receives 4.10 inches of precipitation in the month of May, but in 1943 it had 8.75 inches that month—more than twice the average rainfall. Rainy weather prevailed throughout the month and so retarded field work that only about 15 percent of the corn had been planted at the end of May, when planting is usually completed.

With a tractor and power equipment 3 acres of land can be prepared and planted to corn during the time that 1 acre is planted with work animals. If the tractor is put on a 24-hour schedule, which is not feasible with work animals, the preparation and planting job can be done seven times as fast as with animal power. By utilizing all available mechanical power and equipment (sometimes on a 24-hour schedule) farmers were able to complete their preparation of land and planting of corn in record time during the early days of June, in 1943. It was estimated that 85 percent of the Illinois corn acreage was planted by June 15.

If tractor power had not been available some of the corn would never have been planted because the job could not have been done before it was too late to obtain a crop that year. Tractor power saved the day and, with favorable weather during the rest of the season, Illinois produced 419,000,000 bushels of corn, one of the largest crops for that State on record.

Weather conditions similar to those experienced in 1943 prevailed in many areas of the Corn Belt in 1944 and again in 1945, yet the yields were large. The late wet spring in 1947 delayed corn planting, but this was followed with a late-season drought, and the combined result was a short corn crop. It is apparent that the contribution to production of greater timeliness is one of the real benefits of mechanization, even though it can scarcely be measured in quantitative terms. But the experience in 1947 also indicates that it does not furnish adequate insurance against drought.

RATE OF PROGRESS IN MECHANIZATION

The rate of adoption of mechanical power and auxiliary equipment has varied considerably in different periods. The number of tractors on farms increased sharply during World War I, despite the fact that these tractors were cumbersome, slow moving, and expensive to operate (fig. 7, p. 14). They were adapted only for the heavy tillage and harvesting and for belt work. Farmers had to keep work animals for other farm operations, which meant a considerable duplication of power. After 1920, the rate of increase in tractors on farms slowed down for 2 or 3 years, until the general-purpose tractor was introduced. That was suitable for the cultivation of row crops and for other farm tasks as well as for the tillage and harvesting work.

This revolution in tractor design, along with improvements that were made in standard wheel tractors and crawler tractors for heavy operations on large farms, brought a considerable spurt in

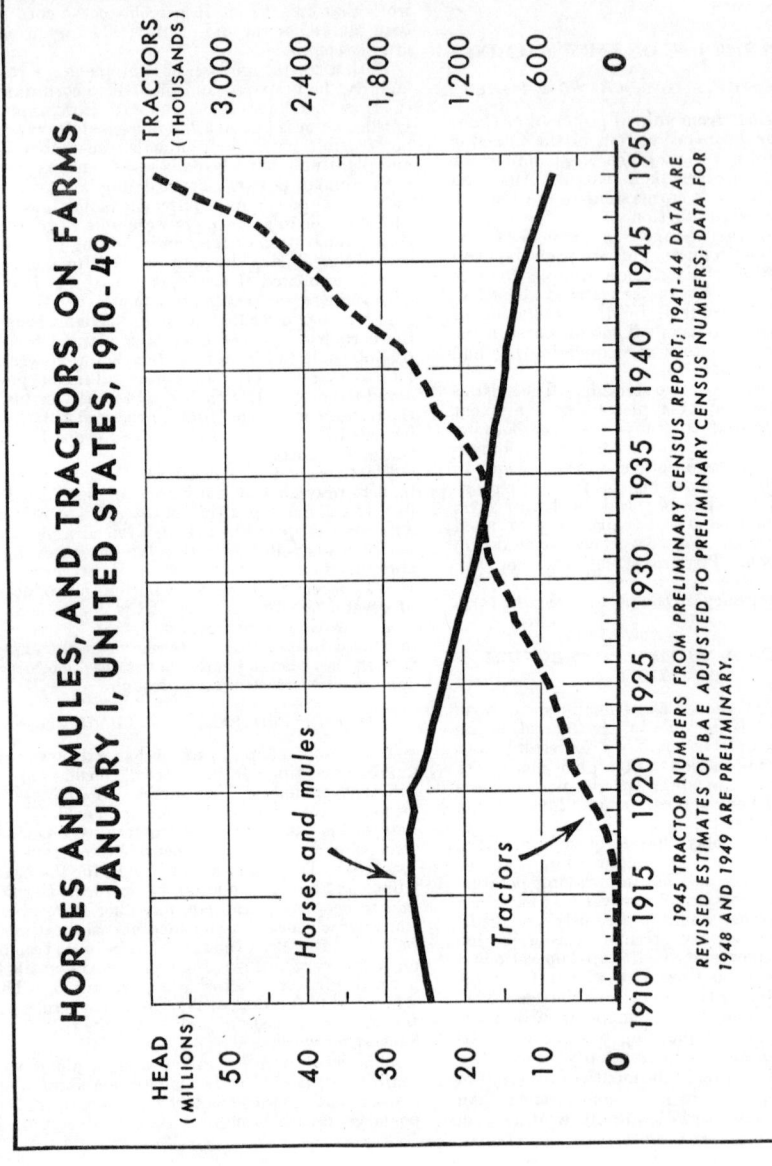

FIGURE 7.—TRACTORS HAVE RAPIDLY BEEN REPLACING ANIMAL POWER FOR FARM PRODUCTION. THE NUMBER OF TRACTORS ON FARMS INCREASED SHARPLY DURING WORLD WAR I AND IN THE 1920'S. THIS TREND WAS INTERRUPTED IN THE DROUGHT AND DEPRESSION YEARS 1931–35, BUT WAS RESUMED IN THE LATE 1930'S AND HAS CONTINUED IN THE 1940'S.

tractor purchases, and a rapid increase in tractors on farms from 1925 through 1930. The numbers increased only slowly from 1930 to 1932 and in the drought and depression years of 1933 and 1934 actually were below the 1932 numbers. With the beginning of the agricultural recovery in 1935, the number increased very rapidly. The sharp upward climb has continued since that time; although war limitations on the manufacture of tractors slowed down the rate of increase, especially from 1943 to 1944. But the numbers on farms increased each year during the war, and on January 1, 1945, they totaled 2,422,000. The estimated number on January 1, 1948 was 3,150,000. The preliminary 1949 estimate is about 3,500,000 tractors on farms.

Introduction of rubber tires for tractors and complementary equipment, during the 1930's, started a rapid and steady advance. It greatly facilitated the use of tractor power for many farm tasks and lowered the cost of tractor operation. Perhaps an even more valuable development in the 1930's was the redesigning of farm machinery for use with tractor power. When tractors were introduced, the machinery that had originally been designed for use with horsepower was adapted to tractors by making special hitches and some other minor changes. The complete changes in design of equipment to facilitate use with tractor power have come only within the last few years.

In summary, the three most important technical developments that stimulated mechanization in the interwar years were: (1) Introduction of the general-purpose tractor adapted for use on smaller farms and for a wide variety of farm jobs, (2) use of rubber tires for tractors and other machines, and (3) design of equipment for use with tractors. These developments accelerated the shift to tractor power and stimulated adoption of combines, corn pickers, and other tractor-drawn machines. Improvements in construction of both tractors and complementary equipment have enabled farmers who are relatively unskilled at mechanical work to operate power equipment without serious disadvantage.

Substitution of tractors for horses and mules has not taken place at uniform rates over the entire country. The rate of adoption has been more rapid in the Corn Belt, the Great Plains, the Mountain, and Pacific States, than in the East and South. Figure 8, page 16, showing the distribution of tractors on farms in 1945, indicates the relatively greatest concentration of tractors in the Midwest and in smaller areas in other parts of the country. Table 1 shows the number of tractors on farms by regions in 1940, 1945, and 1948. The Southern States, excluding Oklahoma and Texas, were far behind the rest of the country in the shift to tractor power up to the beginning of World War II. But purchasing power in the South was built up during that war, and labor shifted heavily to nonfarm work. Both of these developments accelerated the purchase of tractors and complementary equipment. Table 1 shows the consequent upward spurt in tractor numbers in the Southern States from 1940 to 1945 and 1948. In the Southeastern States there were more than 3½ times as many tractors on farms in 1948 as in 1940.

TABLE 1.—*Number of tractors on farms, Jan. 1, 1940, 1945, and May 1, 1948*

Region [1]	Jan. 1, 1940 [2]	Jan. 1, 1945 [2]	May 1, 1948 [3]	Increase 1940–48
	Thousands	Thousands	Thousands	Percent
Northeast	168.0	283.1	387.0	130.3
Corn Belt	463.0	667.2	886.0	91.4
Lake States	252.8	386.0	486.0	92.2
Plains	259.5	349.6	444.0	71.1
Southeast	29.4	66.9	111.8	280.3
Appalachian	52.2	109.9	163.2	212.6
Delta	32.7	65.2	104.0	218.0
Oklahoma and Texas	144.3	232.8	300.0	107.9
Mountain	75.2	120.5	168.0	123.4
Pacific	90.3	140.5	200.0	121.4
United States	1,567.4	2,421.7	3,250.0	107.4

[1] Northeast: New England States, New York, New Jersey, Pennsylvania, Delaware, Maryland. Corn Belt: Ohio, Indiana, Illinois, Iowa, Missouri. Lake States: Michigan, Wisconsin, Minnesota. Plains: North Dakota, South Dakota, Nebraska, Kansas. Southeast: South Carolina, Georgia, Florida, Alabama. Appalachian: West Virginia, Kentucky, Tennessee, North Carolina, Virginia. Delta: Mississippi, Louisiana, Arkansas. Mountain: Montana, Idaho, Wyoming, Colorado, New Mexico, Arizona, Utah, Nevada. Pacific: Washington, Oregon, California.
[2] U. S. Bureau of the Census. United States Census of Agriculture, 1945, 2 v. Washington, D. C.
[3] Bureau of Agricultural Economics.

Continued rapid shift to tractor power can be expected on southern farms. The new small tractors are well adapted to small farms and rolling land. And farmers are learning how to use power equipment to advantage.

The number of tractors on farms will also increase in other areas. As the tractors increase, the over-age horses and mules will be disposed of. There were only 8.3 million horses and mules of all ages on farms January 1, 1949, compared with 14.5 million in 1940. The colts that are being raised are not enough to maintain horse and mule numbers of working age at present levels. Additional land will be released for growing marketable products.

In considering the importance of mechanical power on farms one must not forget the use of automobiles and motortrucks to speed up the

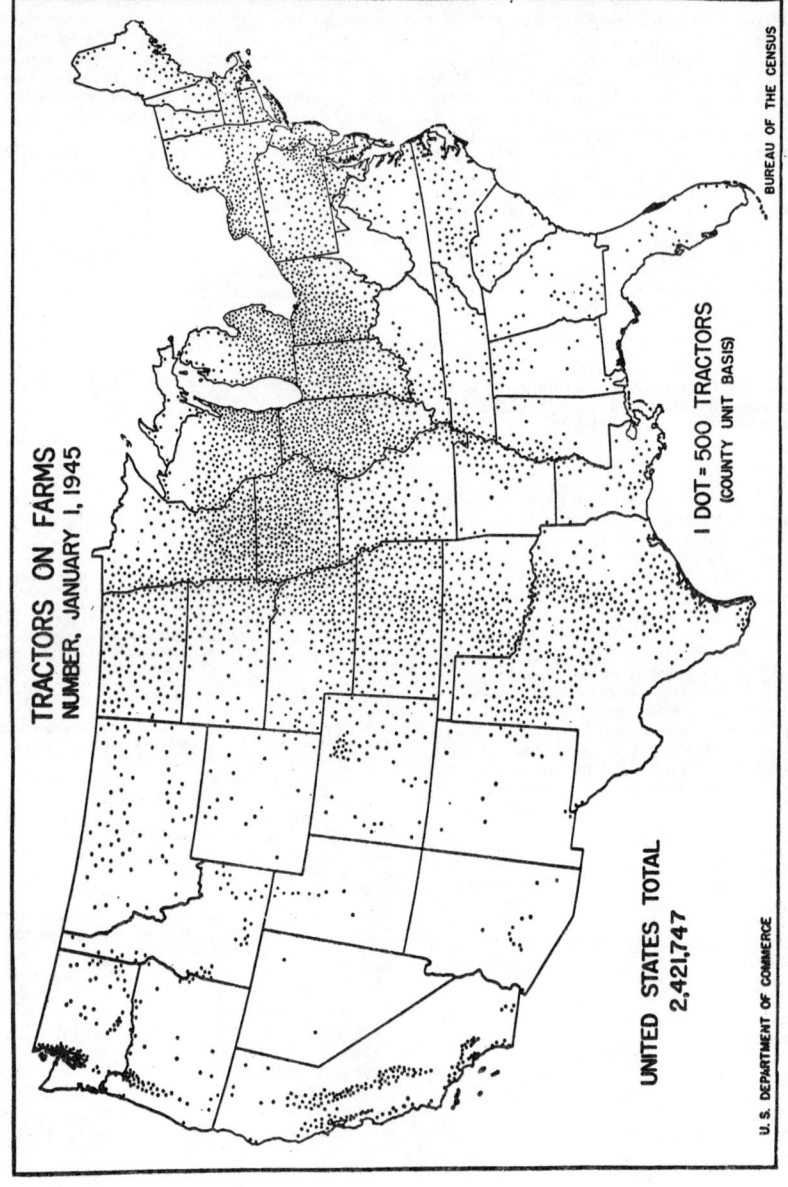

FIGURE 8.—THE GREATEST CONCENTRATION OF TRACTORS IN THE COUNTRY IN 1945 WAS IN THE MIDWEST; SOME SMALLER AREAS IN OTHER PARTS OF THE COUNTRY ALSO SHOWED HEAVY CONCENTRATION OF TRACTORS.

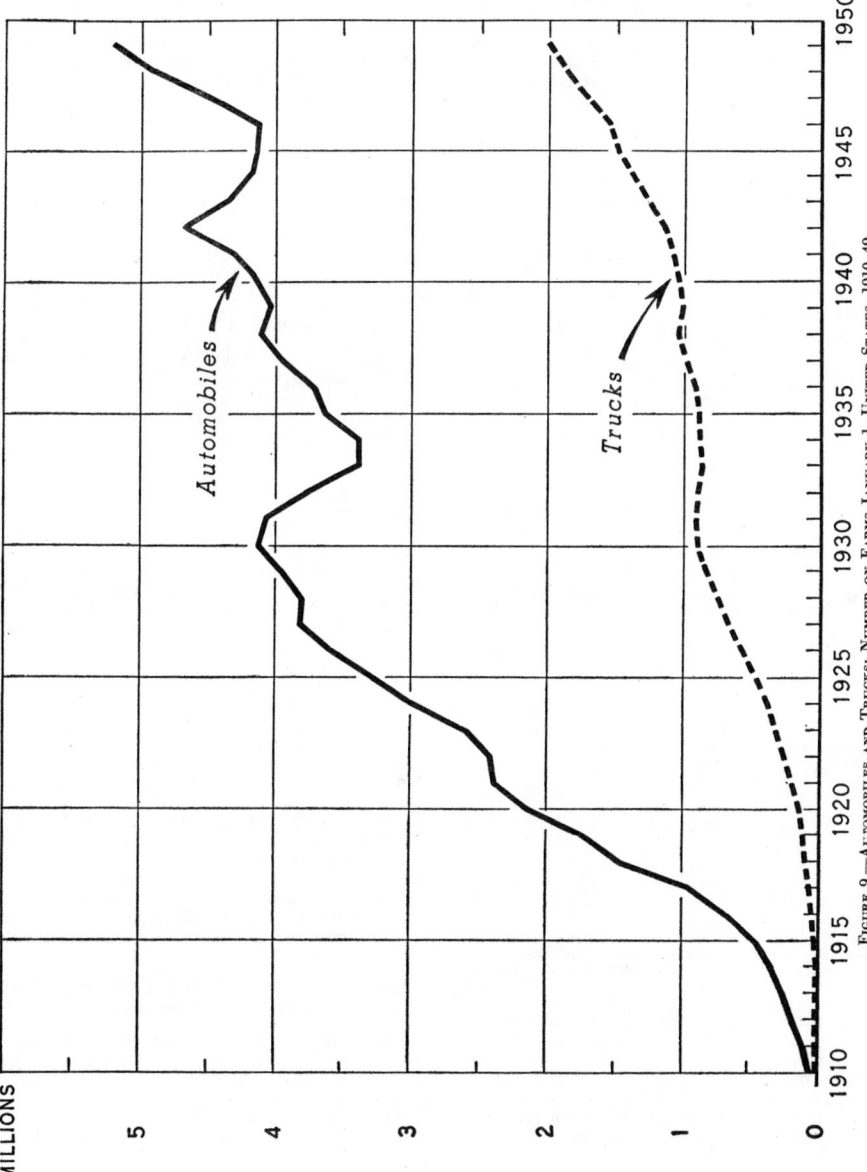

FIGURE 9.—AUTOMOBILES AND TRUCKS: NUMBER ON FARMS JANUARY 1, UNITED STATES, 1910-49.

Automobiles and trucks on farms have increased in the same way as tractors, except for the depression years and for the decline in numbers of automobiles during World War II.

transportation job both on the farms and from farms to market. Figure 9, page 17, shows the increase in numbers of trucks and automobiles, 1910–49. It is apparent that automobiles and trucks on farms have increased in about the same way as tractors, except for larger dips in the depression years of the early 1930's, and for a shrinkage in the numbers of automobiles during the war.

Wartime scarcity of labor stimulated the attempts by farmers to buy more and more labor-saving equipment. In spite of limitations on the manufacture of tractors and other farm machines there were notable increases in labor-saving equipment on farms during the war. For example, the number of tractors on farms increased 57 percent from 1940 to 1945. In the same period the number of grain combines increased 97 percent; corn pickers, 53 percent; and milking machines, 109 percent (table 2). If more new machinery had been available during the war the process of mechanization would have been more complete, especially in the southern States, and farm output would have gone even higher, or more workers could have been released for war industry.

TABLE 2.—*Number of tractors and other specified machines on farms, United States, Jan. 1, 1910–49*

Year	Farm tractors	Farm motor-trucks	Farm auto-mobiles	Grain combines	Corn pickers	Milking machines
	Thousands	Thousands	Thousands	Thousands	Thousands	Thousands
1910	1	0	50	1	------	12
1920	246	139	2,146	4	10	55
1930	920	900	4,135	61	50	100
1940	1,545	1,047	4,144	190	110	175
1941	[1]1,675	[1]1,095	4,330	225	120	210
1942	1,885	1,160	4,670	275	130	255
1943	2,100	1,280	4,350	320	138	275
1944	2,215	1,385	4,185	345	146	300
1945	2,422	1,490	4,152	374	168	365
1946	2,585	1,550	4,150	415	200	465
1947	2,800	1,730	4,520	450	225	580
1948	3,150	1,920	4,930	520	300	640
1949	3,500	2,000	5,250	590	365	685

[1] 1941–44 data are revised estimates of Bureau of Agricultural Economics, adjusted to census numbers; 1945 numbers are from census report; 1946 through 1949 are estimated.

With removal of the wartime limitations on manufacture of machinery farmers stepped up their purchases of all types of farm equipment. And although more machinery was manufactured for domestic use in 1946, 1947, and 1948 than in any prewar year, farmers would have bought a larger volume if it had been available. Tractors, combines, corn pickers, grain elevators, pick-up balers, and heavy disks were among the machines that were in short supply in relation to farmers' desire to buy them during all or part of this period. In the spring of 1949 the supply of most machines was adequate to meet demand at prevailing prices.

Annual purchases of farm machinery are usually closely related to net farm incomes so it is not surprising that demand for new machinery was high in the years 1946–48. Investment in new machinery in good years is one way of building up capital reserves to carry over the years of lower farm incomes. Other accelerating factors in the demand for farm machinery were the high farm wage rates and the favorable experience during the war with the new labor-saving machines.

RAPID PROGRESS IN ELECTRIFICATION

Farm use of central-station electric power expanded rapidly, even during the war. From 1941 to 1945 there were more than 600,000 new installations on farms. In June 1948 about 69 percent of the farms in this country had central-station electric power, compared with 26 percent in 1940 and 9 percent in 1930. Figure 10, page 19, shows the percentage of the total number of farms in each State that had central-station electric power in June 1948. It is expected that many more farms will be electrified within the next 5 years.

Use of electricity in farm production is mainly as a source of stationary power around the farmstead, but the contribution of electric lights and heat to the production of poultry and hogs and to some other enterprises should not be minimized. The use of electric power increases as farmers gain experience with its possibilities. Recent studies indicate a close relation between the number of years a farm has been electrified and the amount of current utilized.

Home uses of electricity are of major importance. For example, in one area in Iowa recently studied about 80 percent of the energy used was for household purposes. Electric power for lighting the home and for cooking, washing, ironing, and other household work, lightens the workload for the housewife. And such home equipment as refrigerators and deep-freeze units supplement farm production by providing for better utilization of farm products, both for home use and for sale.

MECHANICAL-POWER PHASE COMPARED WITH EARLIER DEVELOPMENTS

The impact of mechanical power and complementary equipment on the transformation of agricultural production in the interwar, war, and early postwar years, may well be compared with the agricultural revolution that followed the introduction of improved machinery for use with ani-

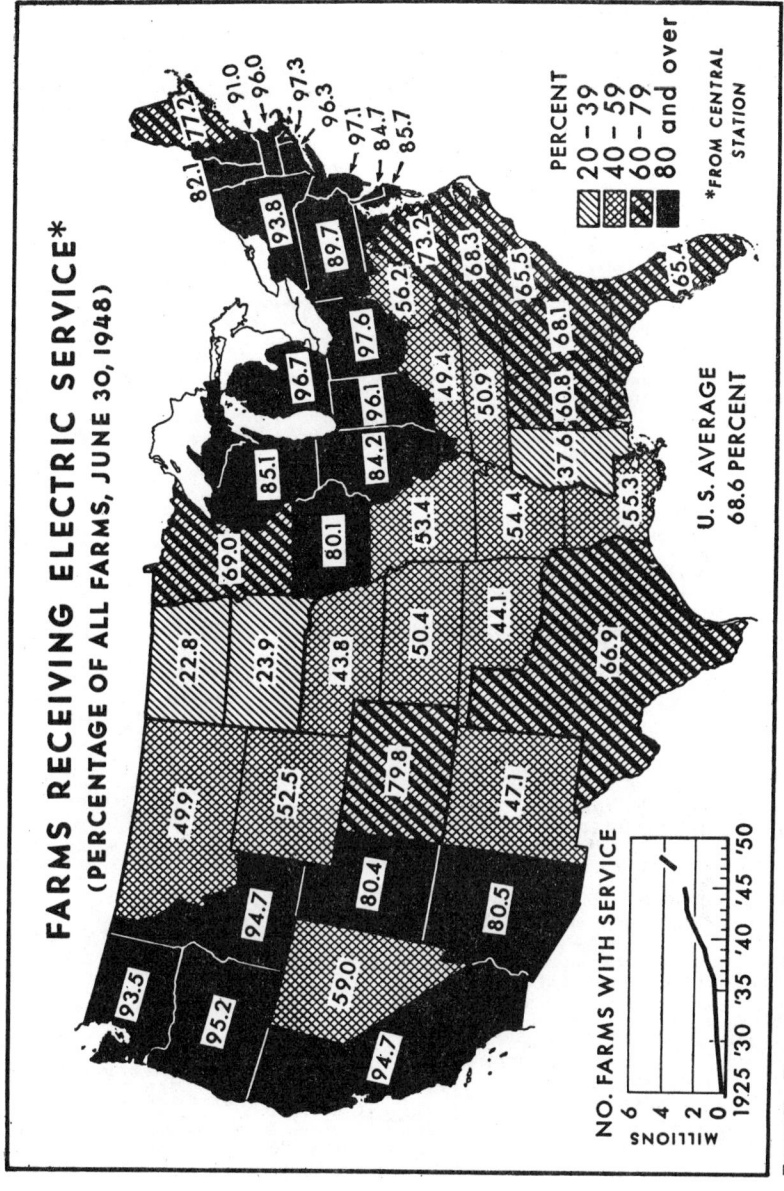

FIGURE 10.—EXTENSION OF CENTRAL-STATION ELECTRIC SERVICE HAS BEEN GREATLY ACCELERATED SINCE 1936. THE ACT CREATING THE RURAL ELECTRIFICATION ADMINISTRATION WAS PASSED IN MAY OF THAT YEAR.

mal power. The steel plow, the mower, the reaper, and later the self binder, were the primary developments in the animal-power phase of mechanization, which began about 1830 and was well stabilized by the beginning of World War I. The mechanical-power phase had then begun, but it did not gain full momentum until the war was over. And the really noteworthy effects have come in recent years.

In fact, the mechanical-power phase of farm mechanization has not yet been stabilized. Wheat production is almost completely mechanized and corn production is well on the way in the main producing areas. Hay harvesting is in a state of flux, with several radically different methods competing for adoption in all the chief hay-producing areas. Mechanization of the cotton harvest is only in its beginning stages, and tobacco is still a hand-labor crop. Considerable progress has been made in labor-saving equipment for the dairy enterprise but relatively less attention has been given to improvements in housing and other equipment for the livestock enterprises than for crop production. Advances in this domain give promise of yielding substantial savings in capital investment as well as in labor.

A look at the new mechanical developments that are already on the horizon leads to the conclusion that the future impacts of mechanization on agricultural production will be just as influential in the decade 1949-58 as were those of the preceding decade. Two of the major developments will be the increase in number of tractors, especially the small sizes fitted with equipment suitable for use on small farms, and the fairly rapid adoption of mechanical cotton pickers and strippers. These developments, along with many others, will have far-reaching repercussions on production and on the number of workers that will be needed in agriculture.

CHANGES IN LAND USE, CONSERVATION, AND FERTILIZER PRACTICES

SHIFTS IN LAND USES

Many changes in the use of cropland and permanent pasture took place between the World Wars. In the early part of this 20-year period the rather extensive abandonment of land in the Eastern States was offset by the development of new lands in the Western States. Total cropland acreage in the United States was a little lower in the early 1920's than in 1919, but it rose slowly from 1927 to 1931 (fig. 11, p. 21). From 1932 to 1939 the trend was downward. Total cropland remained at about the low point reached in 1939 until the war year of 1943; then the total increased to within about 1 percent of the 1928-32 peaks.

Acreages of harvested cropland have been more erratic (fig. 11). The effects of the droughts of 1934 and 1936 are especially apparent. The drought's impact was most severe in the Great Plains and Intermountain States, where much of the sod land that had been broken out in the 1920's was abandoned in the 1930's. Some of it was again brought into use during the war, and by 1947 much more of the land formerly in cultivation was back in crop production. There was also new breaking of native sod lands that over a period of years are best suited for permanent pasture.

Changes in harvested acreages of principal crops in the Northern Great Plains and the Pacific Northwest States are apparent in figure 12, page 22. These States contain the chief wheat-producing areas of the country. It is evident that "the plow that broke the Plains" broke much of it in the decade following World War I. In these States the crop acreage was considerably lower in World War II than in the years preceding the drought and depression of the 1930's. But the crop acreage rose rapidly in the early postwar years.

Changes in the principal uses of cropland from 1928-32 to 1935-39 and to 1944 and 1945 are shown in table 3. These cropland figures include rotation

TABLE 3.—*Changes in the principal uses of cropland in the United States—1928-32, 1935-39, 1944, and 1945*

Use of cropland	Average, 1928-32 [1][2]	Average, 1935-39 [2]	1944 [2]	1945 [2]	Percentage 1945 is of—	
					1928-32	1935-39
	Million acres	Million acres	Million acres	Million acres	Percent	Percent
Intertilled crops [3]	176.6	163.0	168.8	157.6	89	97
Close-growing crops [3]	132.6	133.3	129.8	132.4	100	100
Sod crops [3][4]	77.3	73.5	80.2	82.5	107	112
Total cropland used for crops	386.5	369.8	378.8	372.5	96	101
Summer fallow and idle cropland	41.3	56.9	47.3	54.4	132	96
Total cropland [5]	427.8	426.4	426.1	426.9	100	100

[1] The data on which the 1928-32 estimates are based are less complete than for later periods.
[2] Planted acres so far as available; all others harvested acres.
[3] Adjustments made for multiple use of land by considering first use in the crop year as the primary use.
[4] Including acres in tame hay, hay and cover-crop seeds, and in rotation pasture.
[5] Includes rotation pasture, but does not include wild hay, orchards, vineyards, and farm gardens.

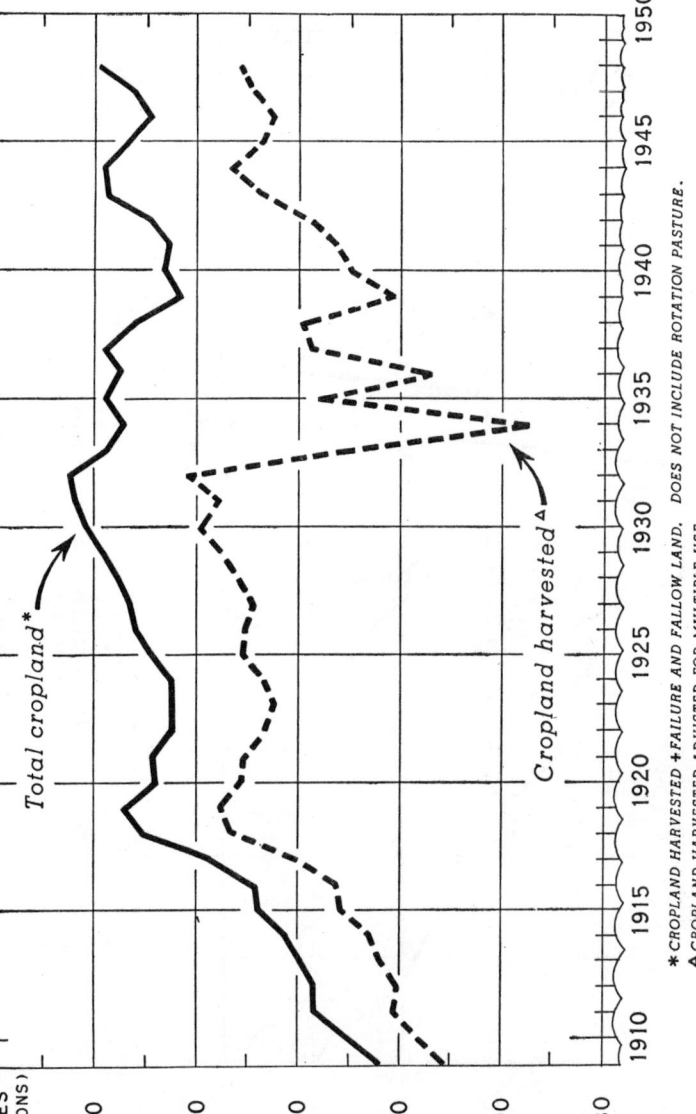

FIGURE 11.—TOTAL CROPLAND AND CROPLAND HARVESTED, UNITED STATES, 1909–48.

Total cropland acreage increased substantially in the decade preceding World War I, and moderately in the first postwar decade. During World War II cropland acreage was approximately the same as in World War I.

22 MISCELLANEOUS PUBLICATION 707, UNITED STATES DEPARTMENT OF AGRICULTURE

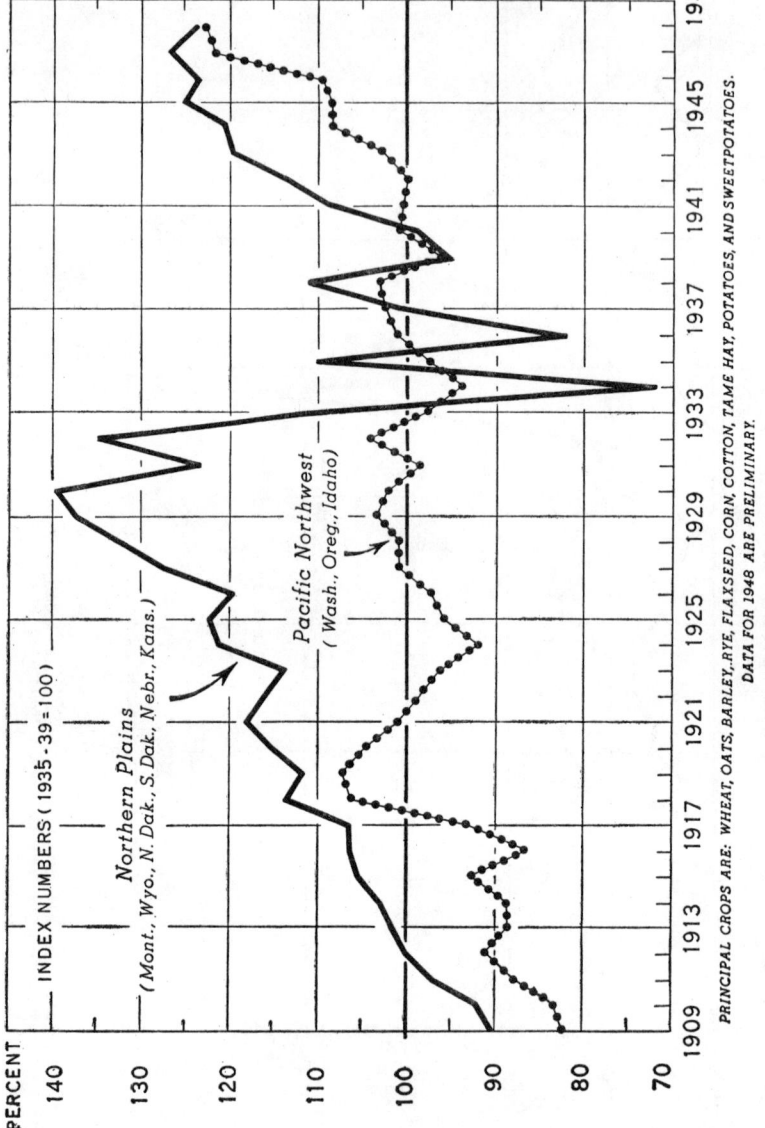

FIGURE 12.—PRINCIPAL CROPS IN NORTHERN PLAINS AND PACIFIC NORTHWEST: HARVESTED ACREAGES, 1909–48.

The acreage of principal crops in the Northern Plains States was higher following World War I than the levels reached at any time during World War II.

pasture but do not include wild hay, orchards, vineyards, and farm gardens. It is apparent that the acreage of intertilled crops decreased, while the close-growing crops held about the same level in the years 1928-32 and 1935-39. The acreage in sod crops decreased from 1928-32 to 1935-39 because of drought and loss of seedings. There was a 4-percent increase in the acreage of intertilled crops from 1935-39 to 1944. The acreage of close-growing crops was decreased, but the sod crops (hay, hay and cover-crop seeds, and rotation pasture) more than held their own.

The acreage in intertilled crops dropped considerably in 1945. There was a steady decrease in intertilled crops in the Southern States during the war, chiefly because of reduced acreage in cotton. But until 1945 that decrease was more than offset by the large increases in corn, soybeans, and other intertilled crops planted in the Corn Belt and Lake States.

The total acreage of land used for crops also dropped from 1944 to 1945. Again the downturn is accounted for chiefly by decreases in the Southern States. The larger wheat acreage is reflected in the somewhat higher acreage of close-growing crops in 1945. By 1947 the planted wheat acreage was 9.1 million acres greater than in 1945, with a large part of the increase occurring in the Great Plains States.

With subsidence of emergency food needs it would be in the interest of permanent agriculture to shift between 8,000,000 and 9,000,000 acres of intertilled crops in the Corn Belt and Lake States into hay and rotation pasture. A part of this shift to sod crops may take place in the next few years, as farmers realize the importance of hay and pasture in a soil-maintaining system of farming. But specific programs may be needed to encourage more rapid progress in that direction. In the humid areas, crop rotations that contain a combination of intertilled, close-growing, and sod crops are more likely to maintain fertility and to sustain high crop yields over a series of years than does single cropping, or a too-great concentration on intertilled crops. Only a part of the evidence of the sacrifice made in postponing crop rotations during wartime is found in the larger acreage of intertilled crops. The five principal Corn Belt States apparently had at least 1.5 times as many acres of intertilled crops that were repeated two or more years on the same land as they had in the immediate prewar years. A cropping program designed to maintain soil resources should reduce considerably this succession of intertilled crops.

In many areas of the Great Plains, the Intermountain, and the Pacific States a part of the cropland acreage should be in summer fallow, as a means of storing moisture and of controlling weed growth. In the wheat areas of the Pacific Northwest the practice of summer fallowing has been followed for more than half a century. In the Palouse area proper, where annual precipitation is 18 inches or more, summer fallowing has not been necessary for moisture storage, and dry field peas are now grown on much of the acreage that formerly would have been summer fallowed. Interest in summer fallowing in the Great Plains was not extensive until the early 1920's. It apparently started in Montana, Wyoming, and Colorado, and extended farther east in the late 1920's and early 1930's, under the pressure of recurring droughts.

Accurate estimates of the acreage of summer fallow by States are not available but table 4 is believed to indicate the general trend. For the seven Western States the fallow acreage has remained substantially the same since 1928-32. Substitution of field peas for fallow in the Pacific Northwest has apparently been offset by moderate increases in wheat acreage in the drier areas where fallowing is necessary to produce a crop. In the Great Plains States, however, summer fallowing reached a peak of 17.4 million acres in 1939—more than six times greater than the average acreage of the years 1928-32. With increased need for wheat, and with more rainfall, fallow acreage dropped by 1944 to around 11,000,000 acres.

TABLE 4.—*Estimated acreage in summer fallow for selected States and periods*

Period	7 Western States [1]	10 Great Plain States [2]	Total 17 States
	Million acres	Million acres	Million acres
1928-32 [3]	5.5	2.8	8.3
1939 [4]	5.4	17.4	22.8
1942 [5]	5.3	14.4	19.7
1943 [5]	5.6	12.2	17.8
1944 [5]	5.5	10.8	16.3

[1] Washington, Oregon, Idaho, California, Nevada, Utah, and Arizona.
[2] Montana, Colorado, Wyoming, New Mexico, North Dakota, South Dakota, Nebraska, Kansas, Oklahoma, and Texas.
[3] Based largely on an unpublished study of the Bureau of Agricultural Economics, 1939.
[4] Agricultural census data for idle and fallow acreage with fallow estimated by applying ratios from rural-carrier survey made for the Bureau of Agricultural Economics.
[5] Estimates of Production Adjustment Committees in each State.

Although no comparable estimates are available for later years, the evidence indicates a further drop in summer fallow of at least 2 million acres. Some reduction was justified as an emergency expedient but widespread resumption of the practice of continuous cropping to wheat would result in a smaller total wheat crop in years of low rainfall. The seedings of wheat for the 1947, 1948, and 1949 harvests were so large in the Great Plains and some other dry land areas that the operators cannot expect to maintain that acreage for several years without reducing their yields. As to permanent soil damage, we have not yet developed crop rotations in the Great Plains that will maintain the soil. Crop farming in that region is therefore to some extent soil mining. But farmers can avoid many of the weather hazards by summer fallowing, stubble mulching, strip cropping, contour farming, and other practices adapted to specific areas. A return to these practices means fewer acres in wheat, but not necessarily fewer bushels of wheat, especially in the drier years.

Since 1936, payments have been made under the Agricultural Conservation Program to cooperating farmers for each acre of protected summer fallow. In some States the payment was made for ordinary summer fallow, provided the surface was ridged or provided stubble mulch was left to prevent erosion. In others, strip fallowing across the direction of prevailing winds or on the contour were the only practices recognized for payment. Most farmers in most wheat-producing areas of limited rainfall are now convinced of the value of these practices.

CONSERVATION PRACTICES

In addition to payments for summer fallow the Agricultural Conservation Program has paid cooperating farmers in all regions for carrying out a wide variety of soil-maintaining or soil-building practices. Among the more important are contour tillage, strip cropping, terracing, and use of green manure and cover crops.

Steady growth is indicated in the adoption of contour operations, the big bulk of which has occurred in the South. Contouring not only lessens the damage from water erosion but is likely to bring increased crop yields, particularly in areas where lack of sufficient moisture often limits the production.

Seeding of green-manure and winter cover crops are practices that protect against erosion losses, contribute in some instances to the supply of available pasture, and increase crop yields through the return of substantial quantities of organic matter to the soil. They are particularly valuable in the more humid areas where a fall-harvested intertilled crop would otherwise leave the land bare, and subject to soil washing in the winter. As yields are maintained or increased through such practices, their rapid extension has no doubt contributed significantly to total agricultural production. The acreage of winter cover crops in the Southern States was about four times larger in 1944 than in prewar years.

Terracing has been done most extensively in the South. The rate of construction of new terraces declined moderately during the war. It is possible for a farmer to build his own terraces with simple equipment, but technical assistance is necessary in laying out the contour lines, and much of the earth moving is done more satisfactorily with heavy mechanized equipment. Using this more costly method may be considered as a capital improvement and, as such, it was delayed by many farmers until after the war. Changes in farming systems in the South which result in a smaller proportion of intertilled crops may permit the development of rotations that will make terracing less necessary in some areas.

Nearly three-fourths of the farms in the United States are now included within the boundaries of soil-conservation districts. By the end of 1948 farmers and technicians had jointly developed long-time conservation plans on about 680,000 of these farms, containing more than 185,000,000 acres. The principal practices applied on these farms are contour cultivation, terracing, strip cropping, crop-residue management, and grass planting. Not all farmers in soil-conservation districts will formulate definite conservation plans. But the farms on which such plans are applied will serve as demonstrations to their neighbors, and conservation practices that prove their worth will spread voluntarily to other farms.

Changes in permanent pasture are much more difficult to trace than changes in cropland. Considerable improvement of permanent-pasture areas was begun in the middle 1930's and has continued. Liming, fertilization, and establishment of new pasture, are the main improvements in the humid areas. In the range areas attention has centered on stockwater development, reseeding, and rotation grazing.

Relatively favorable weather for forage growth made it possible to sustain a large livestock population on the western ranges during the war and early postwar years. If years of lower precipitation should come it would be necessary to reduce the numbers, but in the areas where major improvements have been carried out the long-time carrying capacity has been increased.

In many humid areas the possibilities for further improvement of permanent pasture are great. It seems probable that much of this improvement will be undertaken in the years ahead—part of it in response to the stimulations furnished by conservation programs.

LIMING AND FERTILIZING PRACTICES

Use of lime and commercial fertilizer was greatly accelerated in the immediate prewar and war years, and has continued to increase. A considerable part of the larger farm production can be attributed to the greater use of these materials. But of about equal importance is the potential contribution of lime and fertilizer to the establishment of stable and soil-maintaining systems of farming.

Information on use of liming materials is not available for the years before 1929, when the tonnage of lime used was about 60 percent of the 1935–39 level. The tonnage used annually was nearly doubled from 1935 to 1936, the year when lime was first included in the Agricultural Conservation Program. Figure 13, page 26, indicates that the use of liming materials in the later war years was more than three times the prewar levels, despite the difficulties in obtaining labor and trucks for crushing and hauling. In 1947 more than four times as much lime was used as was used annually in the years 1935–39. Less lime was used in 1948 mainly because of reduction in the program for conservation materials.

Annual use of lime must be increased even more if all of the land in the humid regions that needs lime for soil-improving crop rotations is to receive an initial application, and if adequate maintenance applications are to be made thereafter. It has been estimated that the tonnage of lime applied annually should be about double the 1947 level, to maintain the soil properly and to facilitate desirable shifts toward more grasses and legumes.[3]

On many soils in the humid regions it is necessary to apply lime in order to get full use of commercial fertilizer, especially the phosphates. And lime and phosphate applications are required for successful stands of the legumes and grasses that are so necessary in a good crop rotation, and for soil maintenance. Some of the increase in use of commercial fertilizer is accounted for by the greater use of phosphates in combination with liming materials for hay and pasture improvement. Nearly all the fertilizer distributed by public agencies has been applied on legumes and on hay and pasture lands. But even the relatively large wartime distribution of fertilizer by public agencies accounted for only 10 percent of the total value of the fertilizer that was used.

The increase in consumption of fertilizer is actually an acceleration of a long-time upward trend that was interrupted in the severe depression of the early 1930's (fig. 14, p. 27). The largest part of the increase has been applied to the cash crops, although a growing proportion of the fertilizer now goes on legumes and grasses.

During the war many farmers learned how to use commercial fertilizer to increase their production. It was used in areas and on crops where it had not been used before. This experience is likely to mean a much higher level of use in the coming years. Lower prices for farm products undoubtedly would mean some decrease in purchases of fertilizer for cash crops, but it does not seem at all probable that sales would drop back to prewar levels.

From the standpoint of maintenance of land resources the Nation is vitally interested in the use of lime and fertilizer for the establishment of crop rotations that contain enough acreages of grasses and legumes. Greatly increased consumption of lime and fertilizer for this purpose would help to achieve more stable systems of farming, and would lessen the emphasis on those staple cash crops which seem likely to press most heavily on market outlets.

State committees that were studying postwar adjustments suggested that, under favorable price conditions, it would pay farmers to use about twice the quantity of plant nutrients in the form of commercial fertilizer that was used in the war year 1944.[4] A much larger proportion of it would be used for small grains other than wheat, however, and for legumes, hay, and pasture. These uses would be more than tripled from their wartime levels. If such a program were carried out it would make a substantial contribution to farming stability, and to maintenance of land resources. But the suggested shift in the use of fertilizer toward much greater use on grassland and legumes would not take place without strong educational and other programs that are designed to accelerate the use of lime and phosphate fertilizer in soil-improving rotations.

In common with most other improved practices, greater use of lime and fertilizer, especially on cash crops, increases the volume of farm products that goes to market, which in turn might reduce the prices of the products that are produced in greater volume. But the farmers who use the fertilizer have lower costs per unit of product and a larger quantity for sale as an offset to lower prices. And it is to be remembered that, so far as the increased use of these materials promotes soil-building rotations, the emphasis is shifted away from the staple cash crops that are likely to be overproduced. Such shifts might actually aid in avoiding market gluts of some products. Moreover, part of the increase in lime and fertilizer constitutes a capital investment in permanent soil im-

---

[3] See U. S. Department of Agriculture, Interbureau Committee on Postwar Programs, and The Land-Grant Colleges. PEACETIME ADJUSTMENTS IN FARMING—POSSIBILITIES UNDER PROSPERITY CONDITIONS. U. S. Dept. Agr. Misc. Pub. 595, 52 pp., illus., 1945 for one estimate of lime needs; a later estimate is somewhat higher.

[4] See publication referred to in Footnote 3. The estimated use of nitrogen fertilizer in 1948 was nearly as large as the profitable use indicated in this report; and farmers would have bought more had it been available.

26   MISCELLANEOUS PUBLICATION 707, UNITED STATES DEPARTMENT OF AGRICULTURE

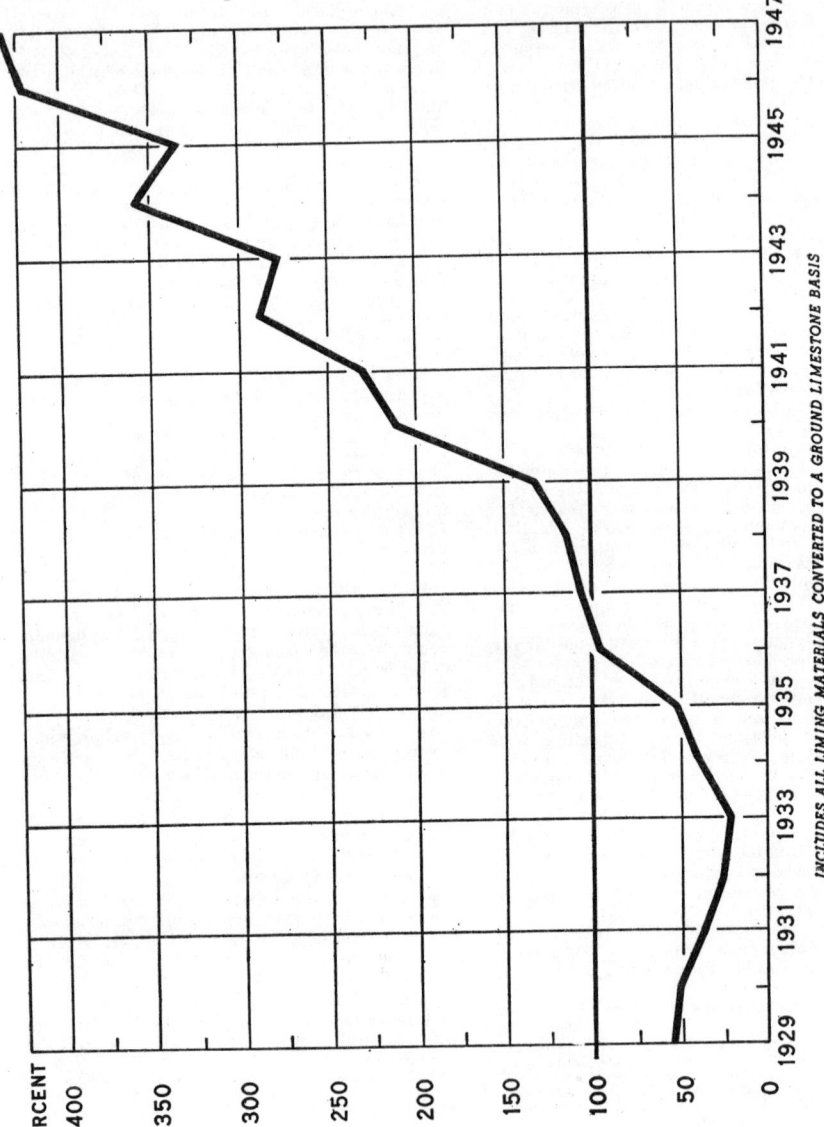

FIGURE 13.—CONSUMPTION OF LIME IN THE UNITED STATES, 1929–47. (INDEX NUMBERS 1935–39=100.)
Application of liming materials had risen to predepression levels by 1935. Most of the increase in the use of lime since 1936 can be attributed to the stimulation provided by its inclusion as a conservation material in the Agricultural Conservation Program.

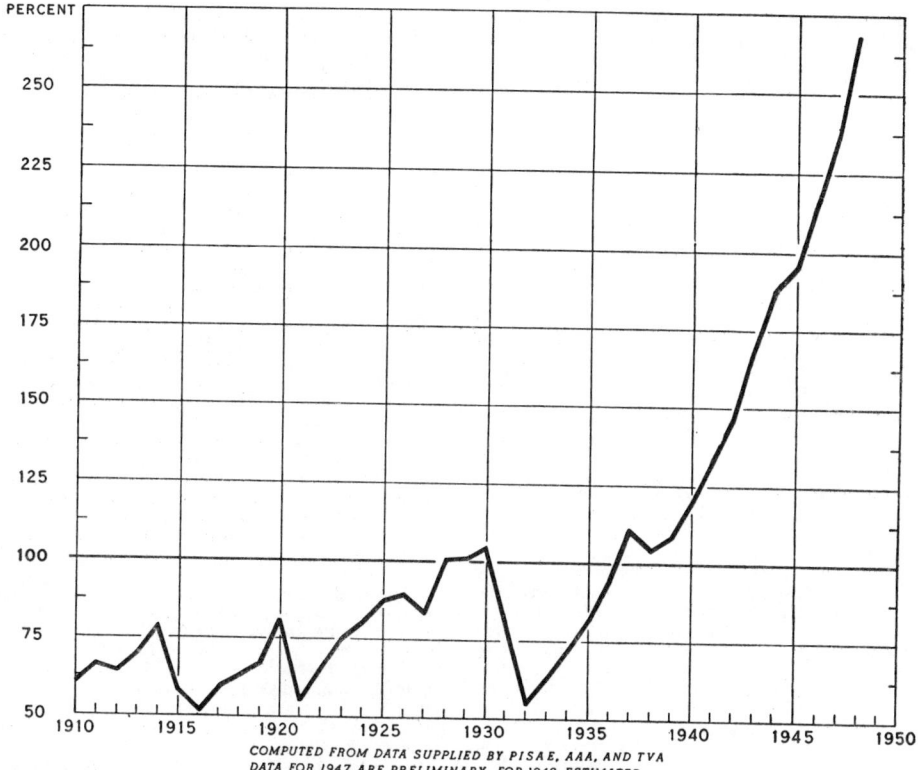

FIGURE 14.—CONSUMPTION OF FERTILIZER IN TERMS OF NITROGEN, PHOSPHORIC ACID, AND POTASH, CONTINENTAL UNITED STATES, 1910–48. (INDEX NUMBERS 1935–39=100.)

Use of nitrogen, phosphoric acid, and potash as fertilizer during World War II reached a level nearly double that of the 1935–39 average, and the use has continued to increase since the end of the war. The highest consumption before 1937 occurred in 1930, when the level reached was 5 percent above the average of the years 1935–39.

provement that is in the interest of future welfare, both for individual citizens and for the Nation as a whole.

CHANGES IN CROPS

The greatest contribution of the changes in crops and varieties made to speed up wartime production was the interwar development of improved seeds that greatly increased the yields per acre of our most important crops. Hybrid seed corn is an outstanding example.

SPECIAL WAR CROPS

During the war the most significant expansion of the strictly war crops was in hemp for hard fiber. Its harvested acreage was increased from 1,248 acres in 1935–39, grown mainly in Wisconsin and Kentucky, to 146,200 acres in 1943 when hemp was grown in several Midwestern States. A Government corporation was organized to contract for acreage in suitable areas, and to construct and operate hemp-processing plants in the new producing localities. Different kinds of vegetable seeds, mung beans, castor beans, and some pharmaceutical plants were grown to alleviate or to protect against war shortages. These were important war crops. But they were concentrated in small areas, they occupied only small acreages, and they constituted a rather insignificant part of the total volume of agricultural production.

INCREASES IN OIL CROPS

On the other hand, large contributions to wartime production were made by stepping up production of some crops that were of minor importance in the interwar years. Soybeans were outstanding among this group. They are one of the oldest of the cultivated crops but their production in this country has occurred mainly in the twentieth century, and commercial production of soybeans has been developed mostly since 1920. The acreage grown for all purposes increased from about 50,000 in 1907 to 460,000 acres in 1917. But in the latter year the output of most of the acreage harvested for beans was used for seed. The acreage harvested for beans to be used both for seed and for crushing expanded from 448,000 in 1924 to 10.2 million in 1944. The acreage harvested for beans in 1944 was 237 percent of the acreage in 1939. About half of this increase represented a shift from harvesting soybeans for hay to harvesting for beans.

Most of the increase in soybean acreage took place in the Corn Belt. Large percentage increases occurred also in the Mississippi Delta and in the Atlantic Seaboard States of North Carolina, Virginia, Maryland, and Delaware. A somewhat reduced acreage in 1946 was followed by a record acreage in 1947. Although the 1948 acreage was smaller than in 1947 it was still at wartime levels.

In the Corn Belt the crops replaced by soybeans were mostly small grains, hay, and rotation pasture. During the interwar years some of the land previously devoted to corn had been shifted to soybeans, but during World War II corn and soybeans increased concurrently. On the whole, soybeans have replaced crops that have lower volumes of output per acre, so the shift has added to the total volume of production.

Figure 15, page 29, shows the growth in acreage of soybeans harvested for beans from 1924 to 1948. It brings out the preeminence of the Corn Belt in the soybean enterprise. But the acreage trends are only part of the story. The trend in yield per acre has been upward, and by far the highest yields have been obtained in the Corn Belt. Thus in the years 1940–44 the five Corn Belt States had 83 percent of the acreage harvested for beans and 88 percent of the total production.

New varieties of soybeans, especially the Lincoln, give promise of further increases in yield per acre within the next few years. The levels of wartime acreage are not likely to be maintained as other sources of oil become more readily available, but it seems probable that both acreage and production will remain at much higher levels than they reached before the war. Soybeans in the Corn Belt have a nearby market for meal, which usually equals or exceeds the value of the oil, so the oil may become the byproduct from soybeans grown chiefly to supply high-protein concentrates. On this basis they can compete more readily with other sources of edible oil.

Peanuts were given special emphasis during the war, and they made a significant contribution to the wartime food supply. Only relatively small proportions of the peanuts have been crushed for oil, however; most were used for nuts or in other direct food products as in peanut butter.

Figure 16, page 30, shows the acreage of peanuts grown for all purposes, the acreage picked and threshed, the yield per acre, and production for the years 1909–48. The acreage of peanuts picked and threshed nearly doubled from the prewar years 1935–39 to 1944. But as the acreage expanded, into new areas and on farms of new growers in old areas, the average yield per acre decreased. In many areas new producers obtained relatively low yields until they had become familiar with the crop. In the Oklahoma-Texas area the acreage was 3.8 times the prewar levels in 1943 but in 1944 it dropped back to 3.1 times the average acreage in 1935–39.

The market for edible peanuts may remain relatively high in the postwar years. But as an oil crop, peanuts are likely to face keen competition that can be met only if prices for oil uses are in line with those for comparable oils. Perhaps more mechanized practices of production can help to give competitive strength to peanut production. But if peanut acreage is maintained at high levels, a large proportion of the total would go into other uses than edible peanuts. Perhaps more of the crop will be hogged off as time goes by.

Flaxseed was the third oil crop that was greatly expanded during the war (fig. 17, p. 31). The 6.2 million acres planted in the peak year, 1943, was more than three times the average of 1935–39, but the acreage planted in 1944 dropped back to less than half that of the previous year, and in 1945 special acreage payments increased the planted area to 4.0 million acres. In 1947 a support price of $6 per bushel resulted in a planted area of 4.2 million acres. This support level was continued in 1948 when 4.9 million acres were planted.

Most of the flax is produced in the spring-wheat States where it was formerly grown as a new-land crop in Minnesota, North Dakota, South Dakota, and Montana. Weed-free land is needed for successful production, but the new weed sprays have proved reasonably effective in controlling weeds in flax fields. Flax is considered a hazardous crop compared with its alternatives. In the drought

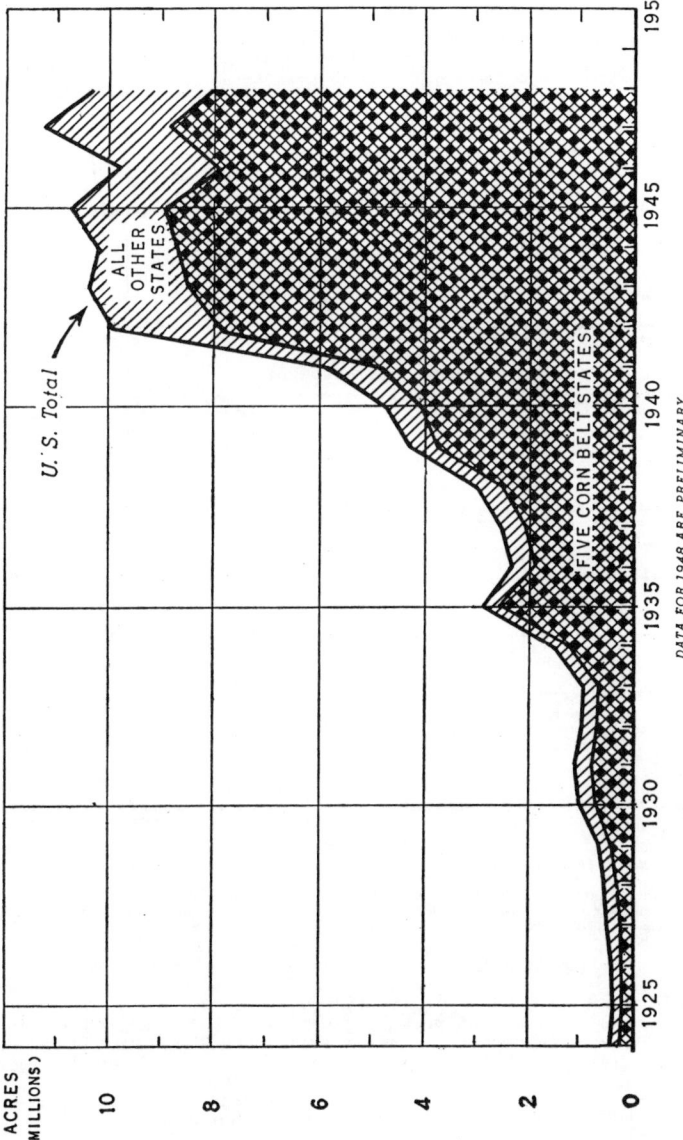

FIGURE 15.—ACREAGE OF SOYBEANS HARVESTED FOR BEANS, UNITED STATES AND FIVE CORN BELT STATES, 1924–48.

The acreage of soybeans harvested for beans in the United States increased gradually from less than a half-million acres in 1924 to about a million acres in the early 1930's, and then rose sharply. The wartime level exceeded 10,000,000 acres.

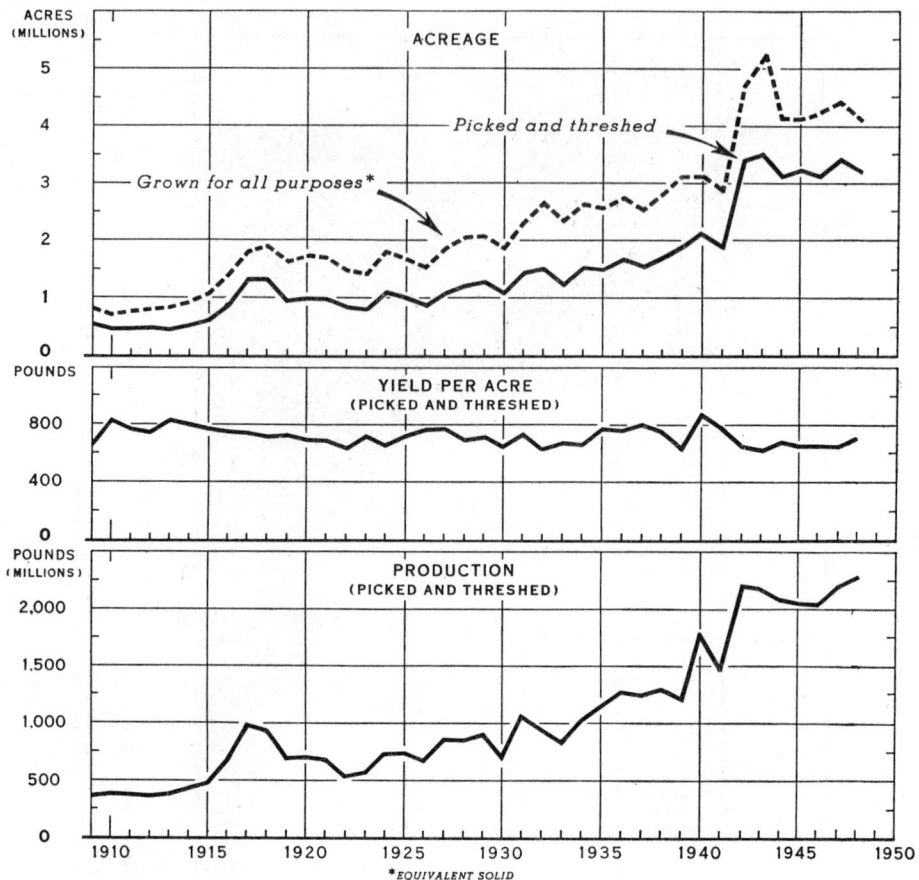

FIGURE 16.—ACREAGE, YIELD, AND PRODUCTION OF PEANUTS, UNITED STATES, 1909-48.

The acreage of peanuts, picked and threshed, nearly doubled during World War II, following the more gradual upward trend of the interwar years. Wartime production did not rise quite so much as acreage because yields declined slightly with expansion into new areas, and on farms of new growers in old areas.

year of 1936 more than 80 percent of the planted acreage was abandoned in the Dakota-Montana area. As compared with the opportunity to grow an unlimited acreage of wheat at loan-rate prices, farmers in the spring-wheat States would have hesitated to take a chance on flax without the special financial inducements that were offered.

Flax yields per harvested acre were higher during the war than in 1935-39. In 1943 and 1944 they were more than 50 percent higher in the Dakota-Montana area. Flaxseed production, on the larger acreage and with the higher yields, in the years 1940 and 1945 ranged from two to nearly five times the 1935-39 level. In 1943 the Dakota-Montana area produced more than eight times as much flax as in 1935-39.

It does not seem likely that the acreage in flax will be maintained at the 1948 figure without spe-

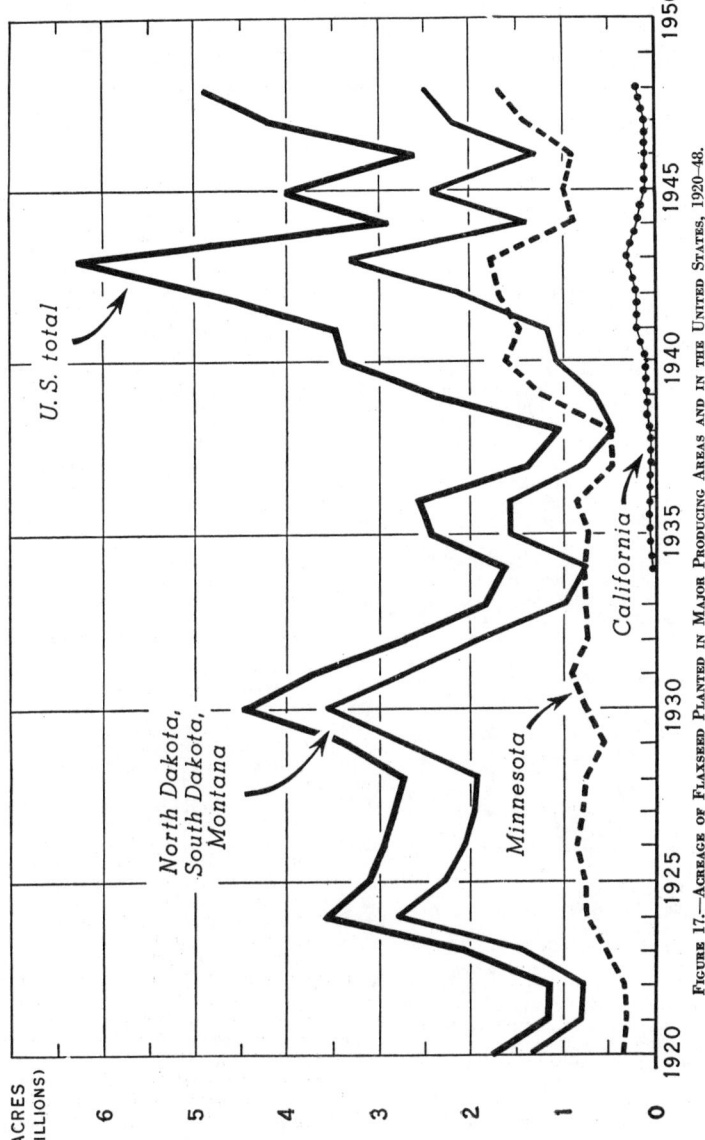

FIGURE 17.—ACREAGE OF FLAXSEED PLANTED IN MAJOR PRODUCING AREAS AND IN THE UNITED STATES, 1920–48.

The acreage of flaxseed is subject to extreme fluctuations. The 1943 acreage was more than three times the average of 1935–39; the acreage fell off sharply in 1944, but it has remained well above the prewar level.

cial price or production incentives. On the other hand, it seems likely to stabilize at levels above the 1935–39 acreage, which was extremely low because of the drought. The acreage planted to flax in the Dakota-Montana area will depend mainly on the comparative returns from wheat and flax.

MORE BEANS AND PEAS

Dry edible beans and dry peas are sources of vegetable proteins that can be substituted, to some extent, for the animal proteins. As they can be readily stored and transported, these direct food crops are well suited to war and relief needs.

The dry-bean acreage increased from the 1.9 million acres planted in 1935–39 to 2.6 million in 1943, which was the peak year. In 1944, the planted acreage dropped to 2.2 million acres, and in 1947 it was below the prewar average of 1.9 million acres. Yields were disappointing in many of the new producing areas, and the competition of other cash crops was too keen to allow the acreage to remain at the high level reached in 1943.

In contrast to beans most of the dry peas are grown as a supplement to other crops rather than in competition with them. Where the annual precipitation is 18 inches or more peas can replace summer fallow on the wheat lands of the Pacific Northwest. The yields of wheat are somewhat lower in a wheat-pea than in a wheat-fallow sequence, but the returns are usually much higher from wheat and peas than from wheat and fallow. This was especially true during the war, when prices were supported at $5.65 per hundredweight for No. 1 peas at country shipping points.

Peas are a highly mechanized crop, and they use about the same machines as wheat. They therefore supplement wheat production with respect to both land and machinery. And, as labor requirements per unit of product are low, the output per unit of additional land, equipment, and labor resources, is high. This is true only in certain areas, however, and there only to the extent that peas can be grown as part of the wheat-pea sequence. Expansion beyond that point means that peas have to be substituted for wheat and grown in succession, or that they replace other crops in other producing areas.

At the beginning of World War II considerable acreage was available for expansion of peas on the basis of supplementary use of resources. Figure 18, page 33, shows how rapidly the acreage of peas expanded during the war. The peak was reached in 1943 with 825,000 acres. This was a much larger acreage than could be grown as a supplementary crop, so some land grew peas in successive years, not only at the expense of wheat but also at a sacrifice of soil maintenance. The crop was grown in some areas that were not well suited to it. The 752,000 acres planted in 1944 represented a closer adjustment of pea acreage to desirable wartime use of resources. In 1948 the planted area was 309,000 acres.

There is no foreseeable domestic food demand for the quantity of peas that have been grown in recent years. They could be grown for a high-protein livestock feed, but the price for them would be much below wartime levels if they were to compete with other protein feeds.

HIGHER YIELDING HAYS

A crop change that developed gradually over the interwar and war years was the shift in the acreage of hay from grasses to the higher yielding legume hays which have a higher protein content and therefore help to balance the livestock ration. Figure 19, page 34, shows the digestible protein available in hay per roughage-consuming unit, by 5-year periods from 1920 to 1944, and for the 4-year period 1945–48. This chart summarizes the changes that have taken place. It indicates that except for the drought years (included in the period 1930–34) the increase has been gradual throughout. The shift toward higher quality and higher yielding legume hays is likely to be accelerated in the postwar years as farmers begin to include more hay and pasture in their crop rotations. The higher protein content of the hay crop will help to balance the ration, and the increased yield of hay will offset at least part of the reduction in volume that otherwise would accompany a smaller acreage of intertilled crops.

ADOPTION OF HYBRID SEED CORN

Development of hybrid seed corn is easily the most important of the interwar and wartime crop improvements. Because corn normally occupies from 25 to 30 percent of the harvested cropland any improvement that greatly increases the yields will naturally have a substantial influence on total production.

Commercial hybrid seed corn was first produced in Connecticut about 1922. Hybrids adapted to the Corn Belt became available in 1929. In 1933 a total of about 143,000 acres was planted with hybrid seed, and in 1948 about 65,097,000 acres. Figure 20, page 35, shows the percentage of the corn acreage planted to hybrids, by years since 1933, in the North Central States and in the United States. Adoption of hybrid seed progressed more rapidly in the North Central States where adapted hybrids were available to farmers earlier, and where corn is the leading farm crop. Adoption is now accelerating in other corn-growing States, especially in the South.

Experience with hybrid seed indicates that acre yields are increased about 20 percent over the yields of open-pollinated varieties. The percent-

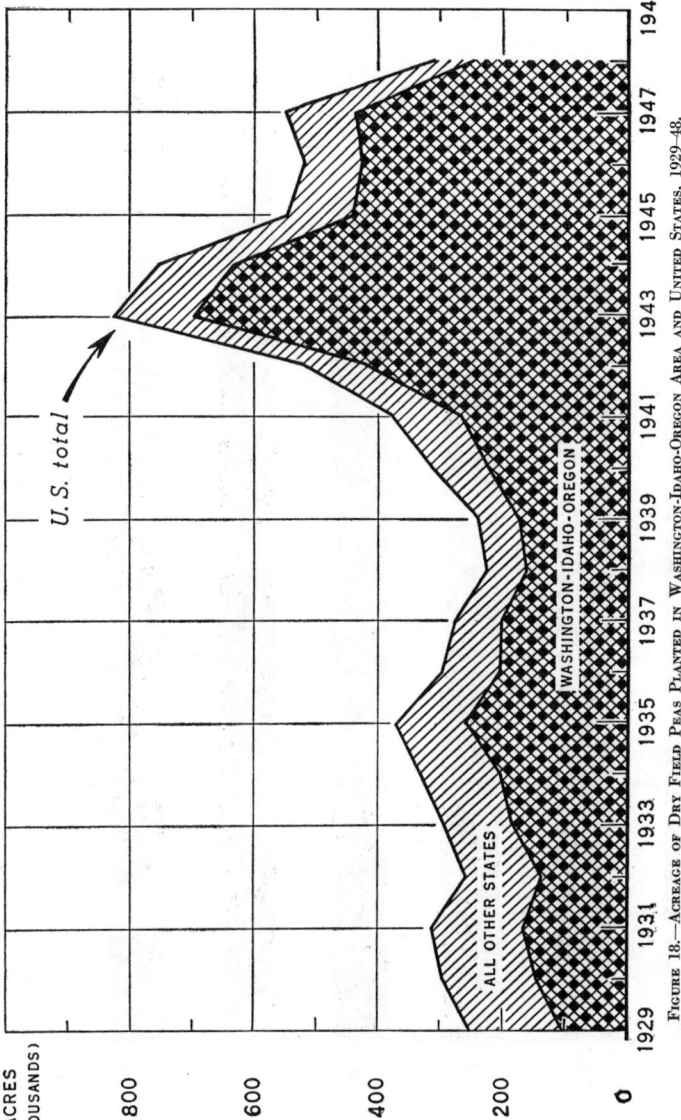

FIGURE 18.—ACREAGE OF DRY FIELD PEAS PLANTED IN WASHINGTON-IDAHO-OREGON AREA AND UNITED STATES, 1929-48.

The acreage of dry field peas rose sharply in the early years of World War II and reached a peak in 1943. The Pacific Northwest States produce the bulk of the output of the United States.

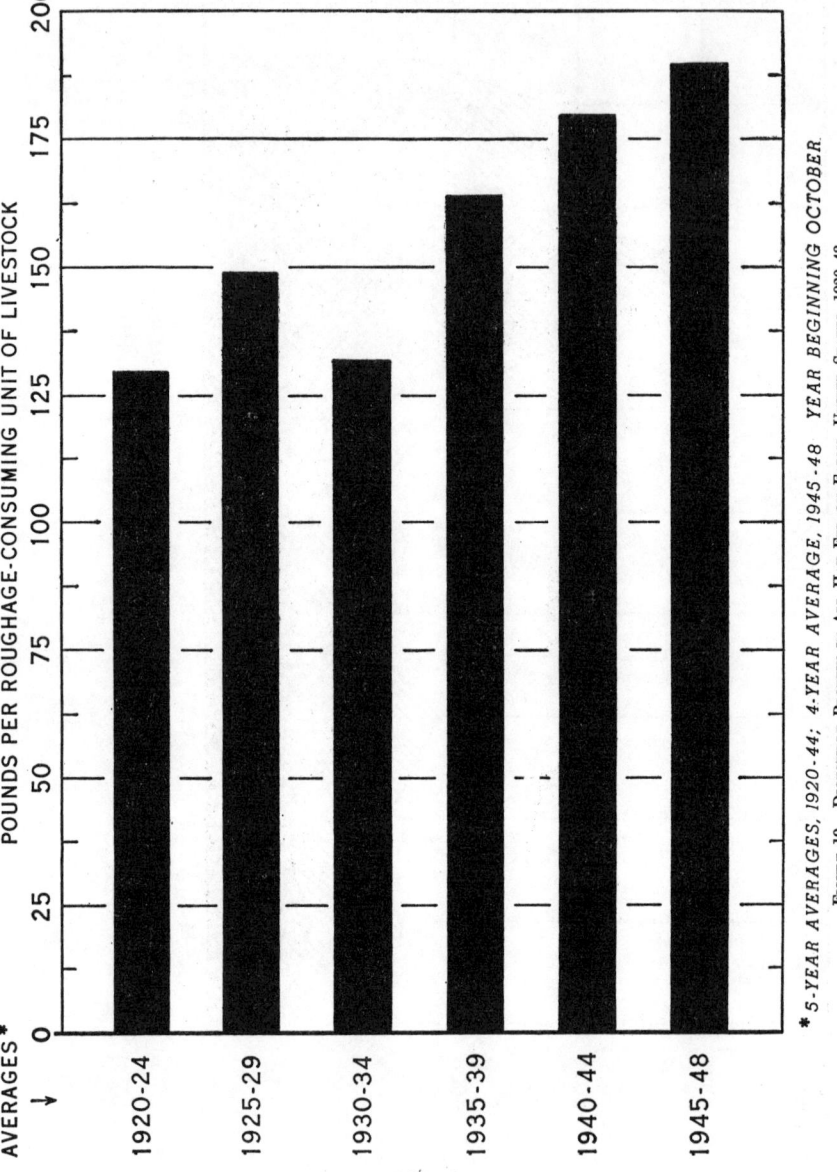

FIGURE 19.—DIGESTIBLE PROTEIN IN ALL HAY FED ON FARMS, UNITED STATES, 1920-48.

Significant changes have taken place in both quantity and quality of the country's hay supply. An average of about 46 percent more digestible protein was available per roughage-consuming unit of livestock in the years 1945-48 than in the years 1920-24. The pronounced shift from grass to legume hays has been the largest contributing factor.

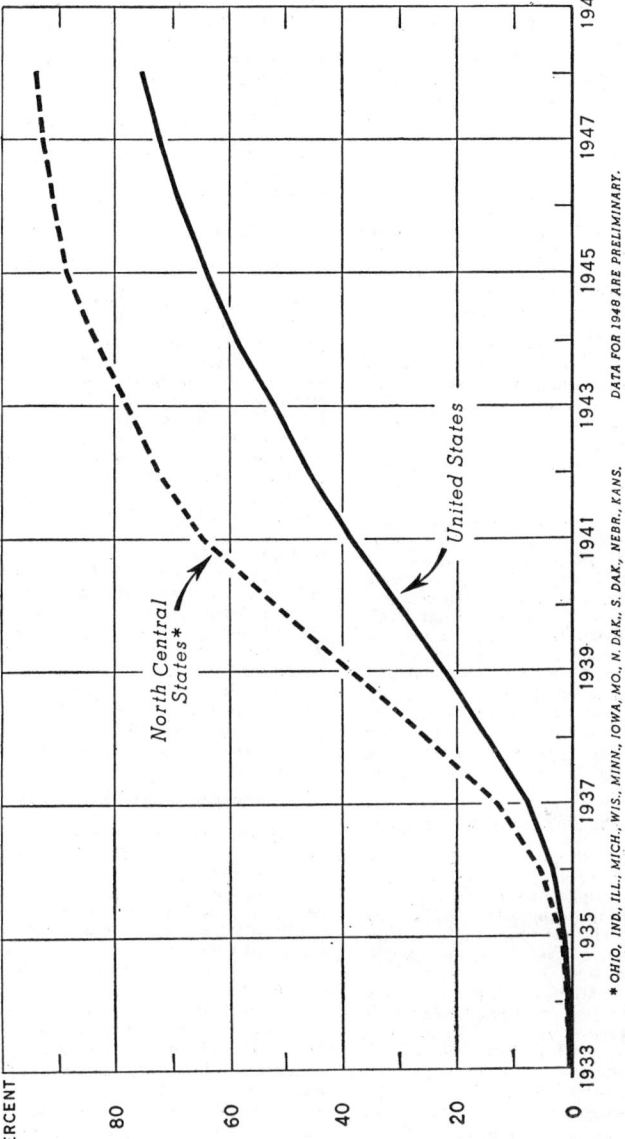

FIGURE 20.—PERCENTAGE OF CORN ACREAGE PLANTED WITH HYBRID SEED, NORTH CENTRAL STATES AND UNITED STATES, 1933–48.

*OHIO, IND., ILL., MICH., WIS., MINN., IOWA, MO., N. DAK., S. DAK., NEBR., KANS.   DATA FOR 1948 ARE PRELIMINARY.

From the merest beginnings in 1933, hybrid seed had replaced open-pollinated varieties on more than three-fourths of the total acreage planted in 1948. The increase has been most rapid in the North Central States, where it was used for planting 94 percent of the acreage in 1948.

age increase is usually about the same on good as on poor land, which tends to give additional advantage to higher yielding areas because expenses do not rise in proportion to the higher yields.

Corn production in 1944 was about a third above the average in 1935–39. The acreage harvested was 1 percent greater, and yields per acre were 31 percent higher (fig. 21). A part of this increase in yields can be attributed to better growing weather than in 1935–39, and a smaller part of it to use of more fertilizer and improvement in cultural practices. But the most effective influence was the greater use of hybrid seed. Assuming an average increase in yields of 20 percent over open-pollinated corn, hybrid seed added 400,000,000 bushels to the 1944 crop. Nearly 16,000,000 more acres would have been required to grow that much corn if open-pollinated seed had been used.

Although the corn crop of 1948 was an all-time record of 3.6 billion bushels, the 1947 crop of 2.4 billion bushels was the smallest since 1936. Power machinery and hybrid seed did wonders in overcoming the handicaps of an unfavorable planting season in 1947, but the dry hot weather which continued through August in the Mississippi Valley was largely responsible for the greatly reduced yields of corn.

The experience with the 1947 corn crop emphasizes that although hybrid seed and power machinery can help to alleviate the conditions of unfavorable weather, they are far from adequate protection against weather hazards. In years of average weather, however, it appears that 3-billion-bushel corn crops will become the rule rather than the exception—even with some contraction of acreage to make room for more hay and pasture. A farmer must bear in mind, of course, that higher yields of corn remove more fertility from the soil, and that if yields are to be maintained he must use more commercial fertilizer, better crop rotations, and more livestock, and make better use of farm manures.

HIGHER YIELDS OF GRAIN SORGHUM AND OATS

Annual production of sorghums for grain in the years 1942–44 was 242 percent of the prewar years 1935–39. The grain sorghums are grown principally in the Great Plains where the yields vary over a wide range, depending upon weather. Figure 22, page 38, indicates that both acreage and production have increased within recent years. Relatively favorable weather, development of new high-yielding varieties that can be harvested with a combine, and the high wartime prices for feed grains, have all contributed to the increased production of grain sorghums. A larger acreage of winter wheat together with low abandonment held down the acreage of grain sorghums in 1945. Less favorable growing weather reduced the yield to 15.1 bushels compared with an average of 17.3 bushels in the years 1940–44. In 1948 production was 131.6 million bushels from 7.3 million acres.

New varieties of oats have resulted in increases somewhat comparable to those of hybrid corn in yields per acre. Better winter varieties adapted for the South have helped to expand acreage and to obtain higher yields in that region. New varieties—as Tama, Boone, Vicland, Marion, and more recently Clinton—adapted to the Northern States, have been grown on a wider scale.

RECORD WHEAT PRODUCTION

Wheat production averaged more than one-fourth higher in 1942–44 than in 1935–39, with a planted acreage only four-fifths as large (fig. 23, p. 39). More favorable growing weather than in those earlier years is of course the outstanding reason. Figure 24, page 40, indicates that a considerable part of the increase in yields during the war represented recovery from the drought yield levels of the 1930's. This is especially evident in the hard-winter and spring-wheat States, which had more than 70 percent of the planted acreage in 1940–44. The successive record wheat crops of 1946 and 1947, and a near record in 1948, are largely attributable to favorable growing conditions which brought high yields on the extensive seeded acreages.

But in addition to this favorable weather there appears to have been an upward trend in yields in the last decade or so that was badly obscured by the drought cycle. It seems reasonable to expect yields higher, by about 2 bushels per planted acre, in the next few years, than the long-time prewar average. Back of this increase are improved varieties, with particular emphasis on disease resistance, also soil- and moisture-conserving practices, and mechanization, which increases the timeliness of operations.

UPWARD TREND IN COTTON YIELDS

Both the acreage and the total production of cotton were lower in the war and the first two postwar years than in 1935–39, but the yield per acre continued the increase that seems to have begun in 1931. The yield receded somewhat in the drought years, but it reached a new peak in 1937, and an all-time record in 1948. Figure 25, page 41, shows the contrasting trends of cotton acreage and yield per acre.

The upward trend in yields of cotton can be attributed mainly to (1) use of more fertilizer, (2) a shift to higher yielding areas with reduction in acreage, (3) careful selection of land within each area and on individual farms, (4) use of improved varieties, and (5) increased use of legumes. As these factors have not operated with equal

CHANGES IN AMERICAN FARMING 37

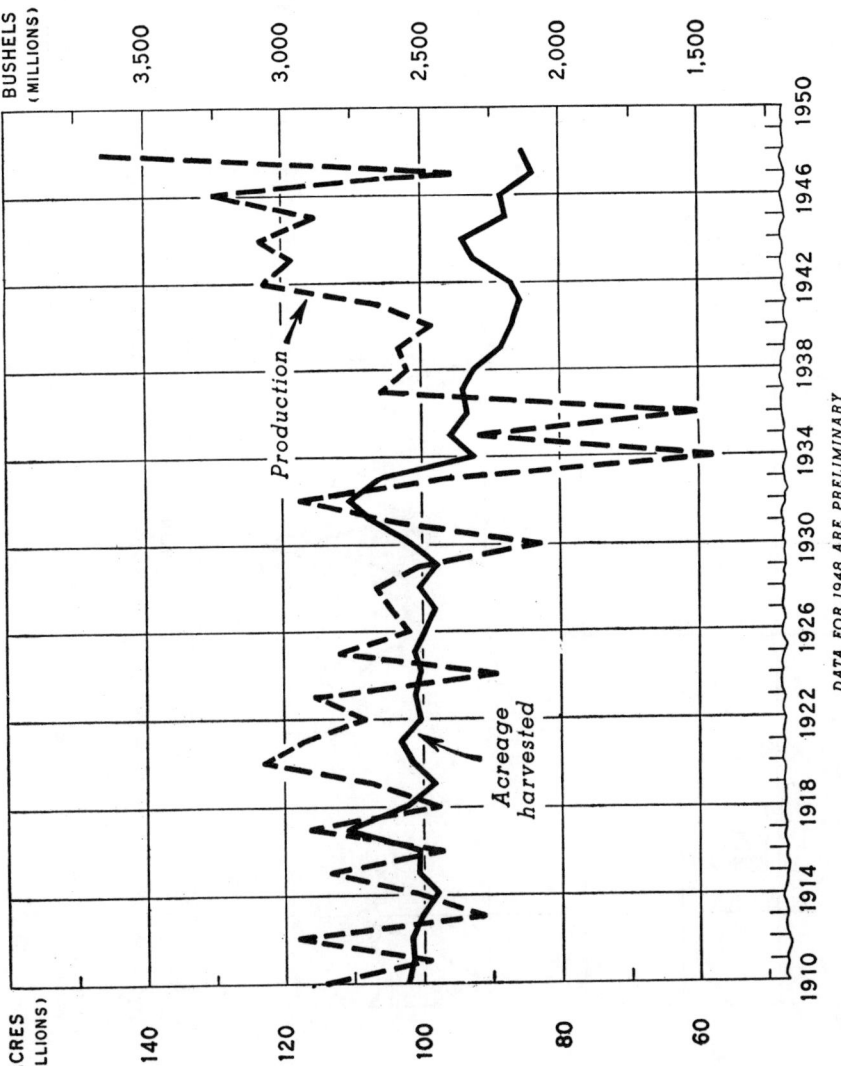

FIGURE 21.—ACREAGE AND PRODUCTION OF CORN IN THE UNITED STATES, 1910–48.

DATA FOR 1948 ARE PRELIMINARY

The acreage of corn harvested in the United States remained approximately stable from 1910 to 1930—at about 100,000,000 acres—except in 1917, when a record acreage of corn was harvested under combined circumstances of World War I and an extraordinary winter-killing of wheat. During the depression—1931, 1932, and 1933—farmers in many areas increased their acreage in an effort to offset low prices. Drought and acreage allotments reduced the acreage before World War II, but the downward trend was reversed in the war years.

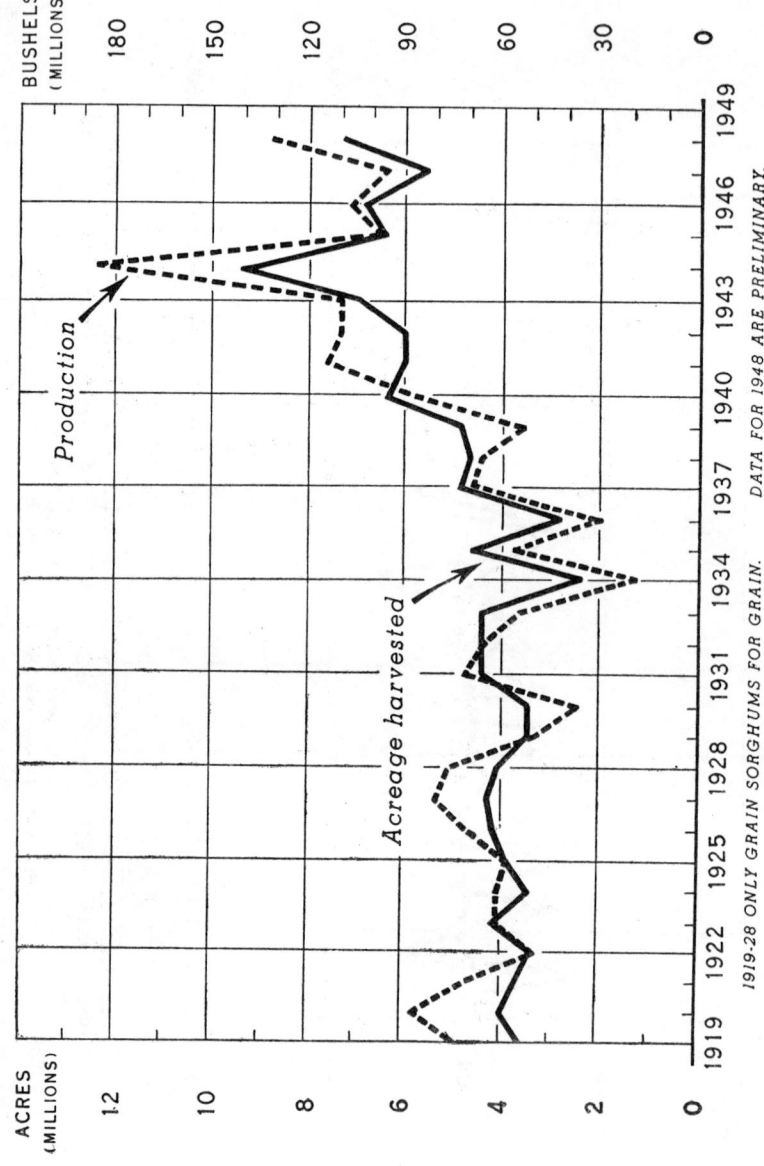

FIGURE 22.—Acreage and Production of Sorghums for Grain, United States, 1919–48.

Between 1919 and 1939 the acreage of sorghums harvested for grain averaged about 4,000,000 acres. Production fluctuated greatly during this period because the crop is grown under very hazardous weather conditions. Both acreage and production since 1939 have been far above the prewar level.

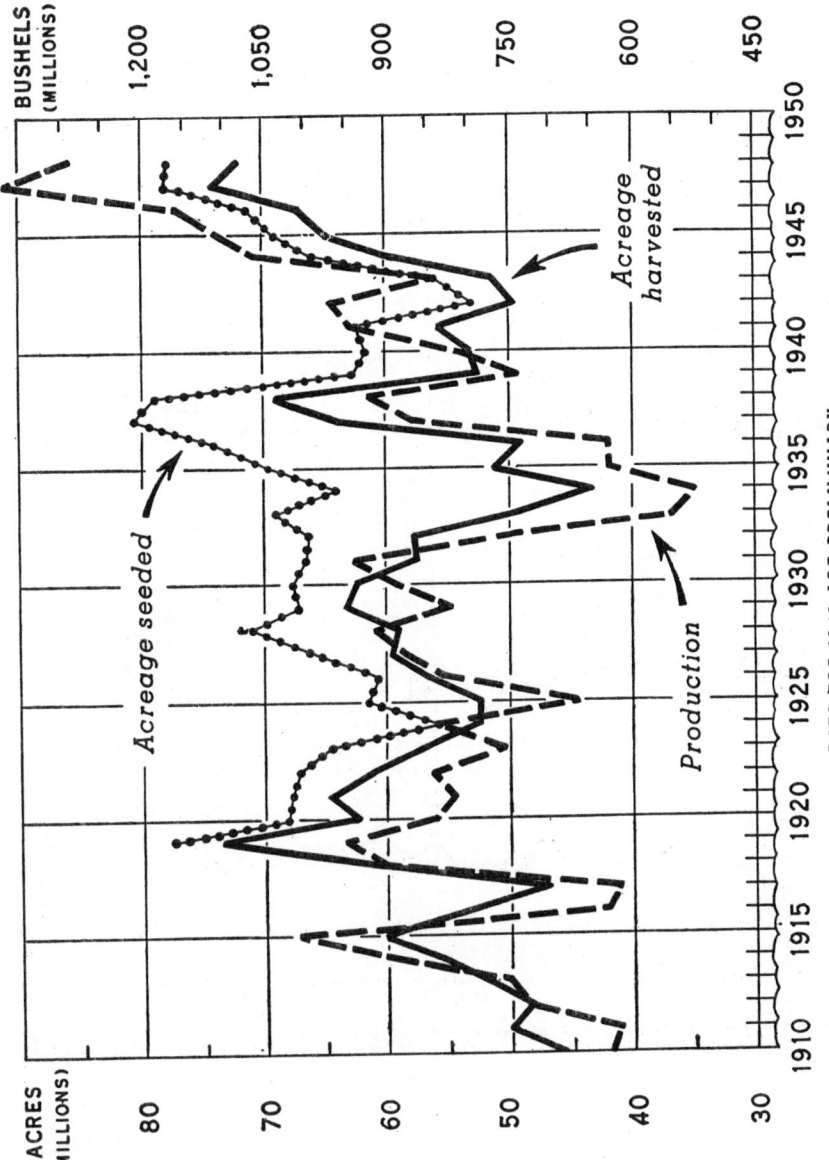

DATA FOR 1948 ARE PRELIMINARY
FIGURE 23.—ALL WHEAT: ACREAGE AND PRODUCTION, UNITED STATES, 1910–48.

Most of the wheat is grown in areas of relatively low rainfall. Therefore both the acreage harvested and production are greatly affected by weather conditions. Acreage and production remained at low levels during the war years, but new records in wheat production were established in the postwar years 1947 and 1948.

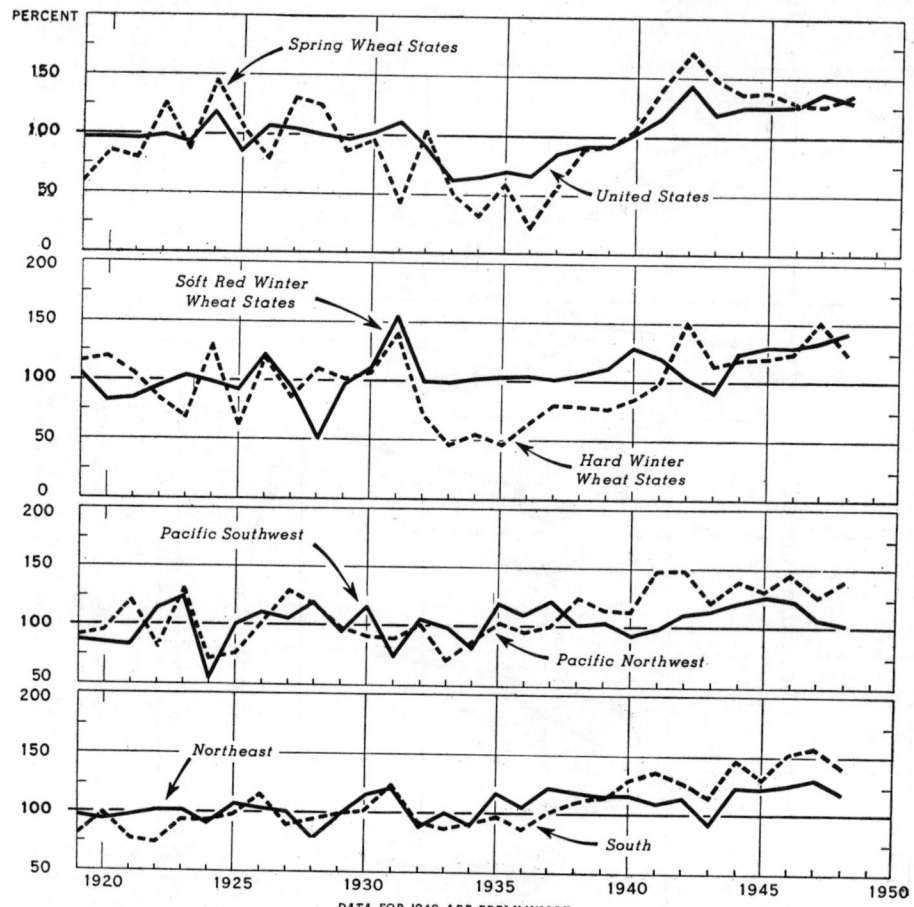

FIGURE 24.—YIELDS OF WHEAT PER ACRE PLANTED, UNITED STATES AND BY GROUPS OF STATES, 1919–48. (INDEX NUMBERS 1923–32=100.)

Yields of wheat have been above the 1923–32 average in most years since 1940, in all regions. Yields in the Pacific Northwest and the Southern States since 1938 have been materially higher than during the 10-year 1923–32 period before the droughts.

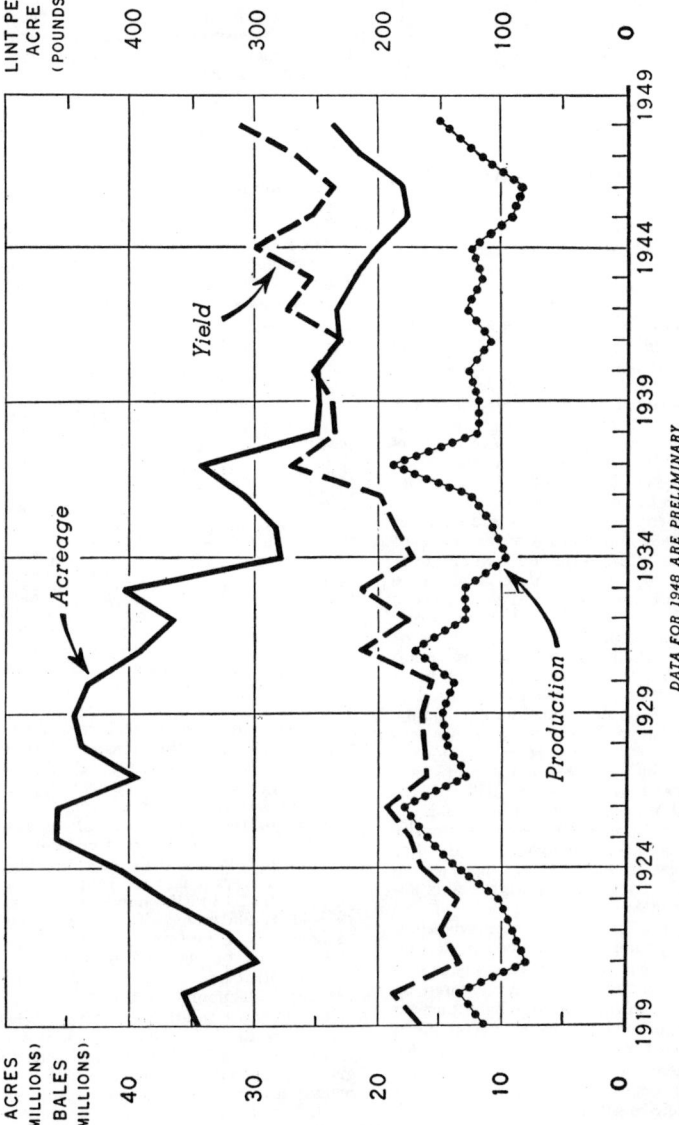

FIGURE 25.—ACREAGE OF COTTON IN CULTIVATION JULY 1, YIELD AND PRODUCTION, UNITED STATES, 1919–48.

DATA FOR 1948 ARE PRELIMINARY

Cotton acreage was increasing from 1921 to 1926, when nearly 46,000,000 acres were in cultivation. Since 1926 the general trend has been downward. Production shows a smaller downward tendency than acreage because the trend of yields has been upward.

effectiveness in all cotton-producing areas, there has been a considerable shift in production, as between areas. This is brought out in figure 26, page 43, which indicates that the Irrigated, the Delta, and the High Plains areas have increased their production of cotton above their 1928–32 levels. In some other areas, as in the Texas Blacklands, production is now much lower than the 1928–32 level.

POTATO YIELDS AT NEW LEVELS

Other crop changes that have influenced the level of production materially are the higher yields per acre of white potatoes, the increase in both acreage and yields of fresh and processing vegetables, and the larger acreage and greater bearing surface of fruit trees.

Changes in yield per acre of white potatoes harvested in selected States and for the United States, from 1919 to 1947, are shown in figure 27, page 44. The upward trend in yields is pronounced in nearly all the chief commercial States. The United States average shows a gradual upward climb; it was interrupted for several years in the early 1930's but reached new highs in the early 1940's. The yield in 1948 of 212 bushels per acre harvested was the highest on record.

There are two principal reasons for the higher yields of potatoes. Adoption of a whole group of improved practices is one—raising higher yielding varieties, use of more fertilizer, and more effective control of insects and diseases. The second reason is that these improvements, combined with mechanization, have pushed more of the production into the hands of specialized commercial growers who use the new methods on large acreages, in areas that are especially adapted to potatoes. This has brought a considerable change in the location of the production both within several of the chief producing States and among the different States. The harvested acreage of potatoes has gone generally upward since 1919, in California, Idaho, and Maine. On the other hand, the acreage in the Lake States (Minnesota, Wisconsin, and Michigan) has had a downward trend. The most rapid increase in recent years has been in Kern County, Calif., where very high yields are obtained under irrigation.

Gradual concentration of potato production, principally in the hands of specialized producers who are located in the most favorable areas, is still under way. This trend has been accelerated by the price-support programs and may be influenced by changes in support prices. But it seems probable that if weather conditions are favorable, new records of yield may be established as the large-scale, specialized producers make further improvements in their methods.

EXPANSION OF VEGETABLE PRODUCTION

Acreages devoted to commercial truck crops increased sharply between the two wars (table 5). Improvement in techniques of production, better knowledge of disease and insect control, and better methods of marketing and distribution, gave producers added incentives on the production side. Greater appreciation by consumers of the nutritional value of vegetables in the diet contributed to an expanding consumer demand and provided a market for a corresponding higher level of production.

TABLE 5.—*Harvested acreage of commercial truck crops for fresh market and for processing, United States, 5-year averages, 1920–44, and annual 1940–45*

Year	For fresh market	For processing	Total
	Thousand acres	Thousand acres	Thousand acres
1920–24	739	741	1,480
1925–29	1,160	1,025	2,185
1930–34	1,600	1,064	2,664
1935–39	1,744	1,386	3,130
1940–44	1,699	1,799	3,498
1940	1,711	1,394	3,105
1941	1,682	1,664	3,346
1942	1,649	1,997	3,646
1943	1,573	1,958	3,531
1944	1,879	1,984	3,863
1945	1,893	1,943	3,836
1946	2,047	2,062	4,109
1947	1,843	1,879	3,722
1948 [1]	1,802	1,710	3,512

[1] Preliminary estimate.

Acreages of fresh-market vegetables expanded more rapidly than the acreages of vegetables for processing between the wars. Production of processed vegetables increased rapidly after 1933, however, partly because of improvements in the freezing of foods and the resulting rapid expansion in demand for frozen vegetables.

Consumption of fresh vegetables, per capita, increased in average figures from 205 pounds in the period 1920–24 to 235 pounds in the 5-year period 1935–39, to 237 pounds in 1943, and to 256 pounds in 1948. Consumption of canned vegetables was 19.6 pounds net canned weight per capita, in 1920–24, 29.9 pounds in 1935–39, 34.5 pounds in 1943, and 38.3 pounds in 1948.

The upward trend in acreage of fresh-market vegetables in the interwar period was reversed from 1940 to 1943. This was influenced in part by growers' difficulties in obtaining farm labor and production supplies, by more emphasis on other needed commodities, by a shift on the part of many

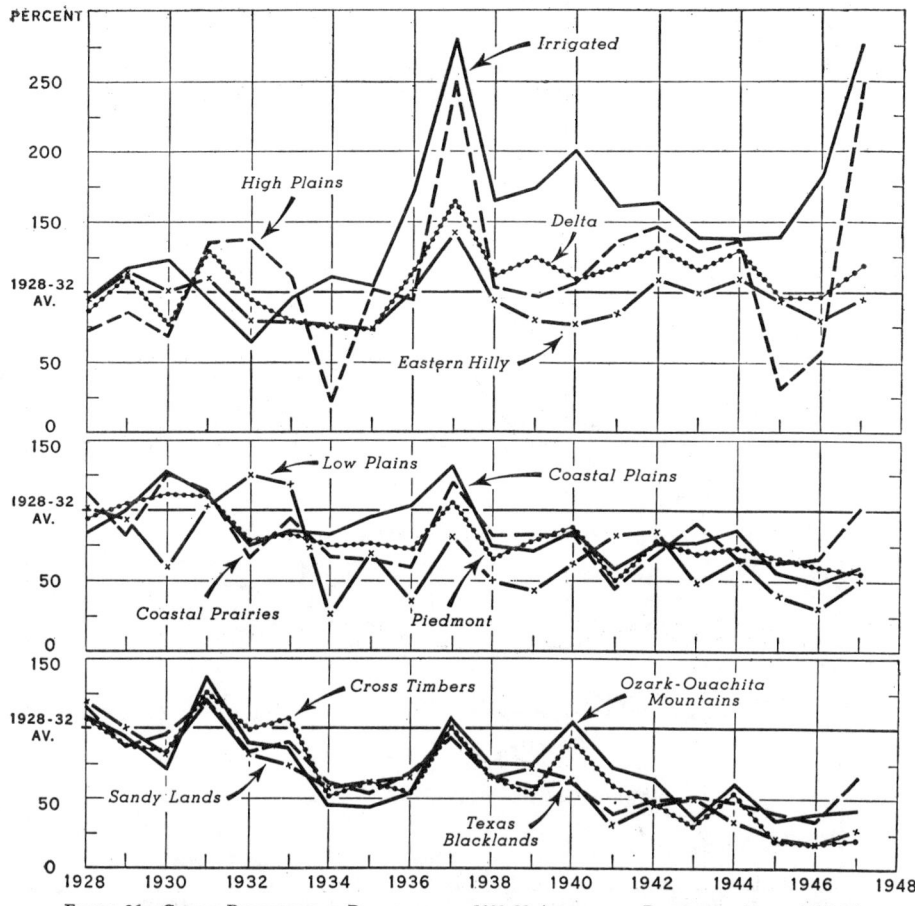

FIGURE 26.—COTTON PRODUCTION AS PERCENTAGE OF 1928–32 AVERAGE, BY PRODUCTION AREAS, 1928–47.

The wartime cotton acreage was below the prewar level in all except the irrigated areas. This resulted in decreased production except in the areas where higher yields have more than offset the smaller acreage.

growers from fresh vegetables to vegetables for processing, and by the victory gardens that displaced some of the commercial market. But a record acreage was planted in 1944, which was exceeded in 1945 and again in 1946. Somewhat smaller acreages were harvested in 1947 and 1948.

In contrast to the downward trend of fresh-market vegetables during the first three war years, the acreage of vegetables for processing increased sharply in 1941 and again in 1942. It was maintained at or about the 1942 level from 1943 to 1946, and was down a little in 1947 and 1948. This means an average about 40 percent above the 1935–39 average. Heavy wartime demands for processed foods for military and lend-lease use provided the basis for this expansion. The major

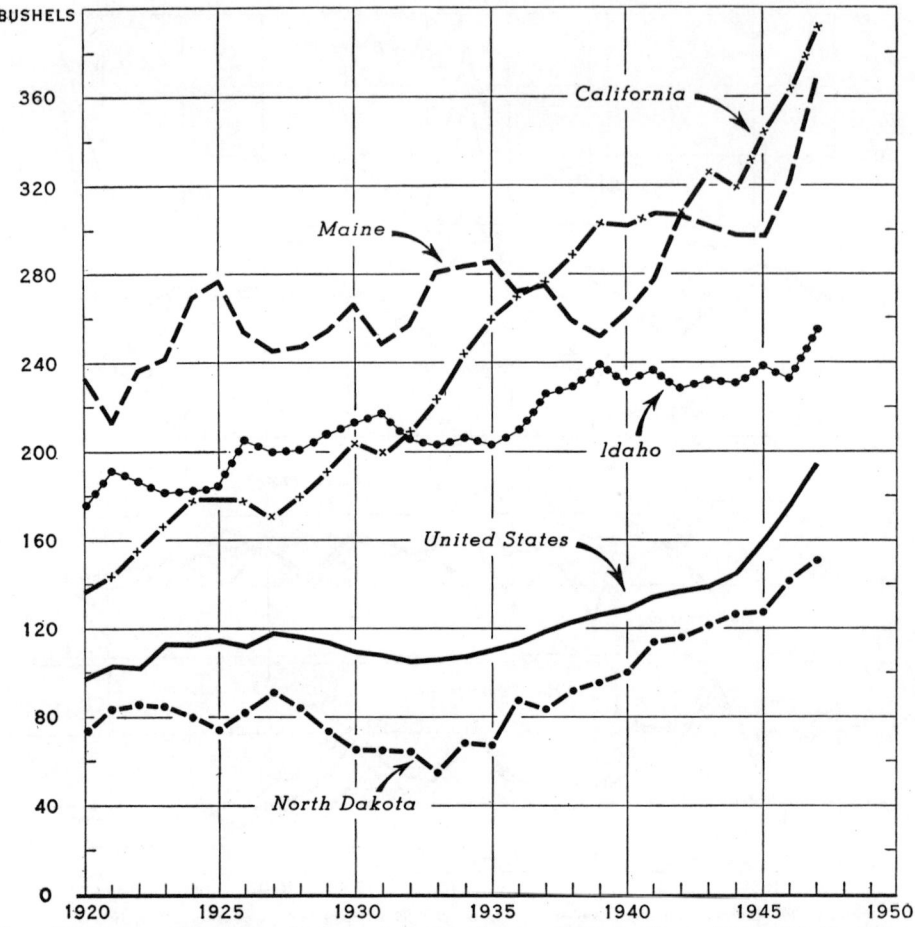

FIGURE 27.—YIELD OF POTATOES PER HARVESTED ACRE, FOUR SELECTED STATES AND UNITED STATES, 1919–48. (3-YEAR MOVING AVERAGE).

Maine, Idaho, and California produce about one-third of the potatoes grown in the United States. They are the only States that consistently show yields above 200 bushels per acre. Production there is concentrated on land best suited to potatoes and in relatively large-scale enterprises. Production in North Dakota has shown a marked shift from farm production for home supply to commercial production in the Red River Valley. The decided upward trend in United States average yields reflects the extension of acreage by commercial growers who apply effective production practices.

processing vegetables were supported at prices favorable to the growers and this encouraged grower interest in meeting the requests for greater production.

The demand for canned vegetables may now slacken considerably, but there is every reason to anticipate an expanding outlet for frozen vegetables which may more than compensate for any decline in the quantity of processing vegetables that goes into cans. The peacetime outlook for all vegetables will be closely related to the level of consumer purchasing power, but a further rise in per capita consumption can be expected.

GROWING IMPORTANCE OF CITRUS FRUITS

Bearing acreage of tree fruits is not susceptible to much change within a few years, so higher demand is reflected mainly in higher prices rather than in increased output. Most commercial acreage must be well cared for at all times, and orchards that are allowed to deteriorate through lack of care usually cannot be restored. More fertilizer and better cultural practices result in higher yields in many orchards, but the scarcity of labor and supplies during the war frequently prevented using these means to increase the output. Changes that occurred during the war, therefore, were usually continuations of trends that were in evidence during interwar years. Strawberries and cranberries are the only important fruit crops in which acreage and production were materially affected by the war. Acreage of these two crops combined declined from 205,000 in 1941 to 107,000 in 1945. Decline in production was even more marked.

Acreage of deciduous fruit trees of bearing age reached its highest peak in the late 1920's and early 1930's. During the last decade the acreages of most of the deciduous fruit trees have declined gradually, but citrus trees have increased. Total production of combined fruits has increased; this is attributed almost entirely to the rapidly expanding production of citrus fruits. Fruits other than citrus have remained at a relatively constant production level during the last 15 years (fig. 28, p. 46).

The striking point in the fruit picture is therefore the change that occurred in citrus production. From a relatively small part of the output, citrus fruits have increased until they constitute almost one-half of total fruit production, on a tonnage basis. Citrus production rose from about 2,000,000 tons in 1929 to 7,100,000 tons in 1943—a threefold increase in 15 years. The trend is still sharply upward. As most of the orange and grapefruit trees have not yet reached full production, with present numbers of trees and with normal care of orchards, the production and yields can be expected to keep on rising through several decades, for the trees are relatively young and they naturally have a long productive life.

Yields of all fruits on a bearing-acreage basis have increased. The higher average age of the trees probably has been the most influential single cause. This holds true even for apples, although little further increase in yields can be expected because of older trees.

A shift to fruit production usually means that a product of higher value per acre is obtained as soon as the orchard is of bearing age, and the output increases as the orchard grows toward maturity. These facts partly account for the large increases in farm production over a period of years in Florida and in the Pacific States.

Because many fruits are still considered to be virtually a luxury food by many consumers, the level of national income will be of even greater concern to growers of fruits than to producers of other agricultural commodities. If costs of both production and marketing can be reduced, and if such improvements are reflected in prices that induce larger consumption, the markets for fruit can be gradually expanded.

CHANGES IN LIVESTOCK PRODUCTION

FACTORS INFLUENCING CHANGES

Three major forces are back of the recent changes in livestock production. They are (1) the shift from animal to mechanical power, (2) the variations in the total feed supply, and (3) the higher production per animal. These forces in turn have been influenced by prices of livestock products and other economic changes.

The decrease in horses and mules of 15,000,000 animal units from 1920 to 1946 released land that could grow feed for an equivalent number of productive livestock (animals and their products that are produced for human use). The saving in grain alone amounted to about 16,000,000 tons in 1946—enough to feed 32,000,000 hogs to market weight.

Year-to-year changes in the total feed supply have been about as influential as the shift to mechanical power in their effects on livestock production for human use. The severe drought years, 1934 and 1936, reduced total feed consumption (feed grains, hay, and pasture) about one-fifth below the 1928–32 average. On the other hand, total feed consumption in the war years 1942–44 averaged about 28 percent above the 1928–32 levels, and 34 percent higher than those for 1935–39. Feed production in 1942–44 increased more than in proportion to the increase in numbers of livestock, which meant that there was more feed available per animal. Total numbers of livestock fed increased 20 percent above 1935–39 but the increase in livestock production was greater than this. Excluding horses and mules the increase in production was about one-third.

Livestock production was reduced in 1946 and again in 1947. The short corn crop of 1947 placed definite limits on expansion in the early part of 1948, but the favorable livestock-feed price ratios resulting from the large feed-grain crop of 1948

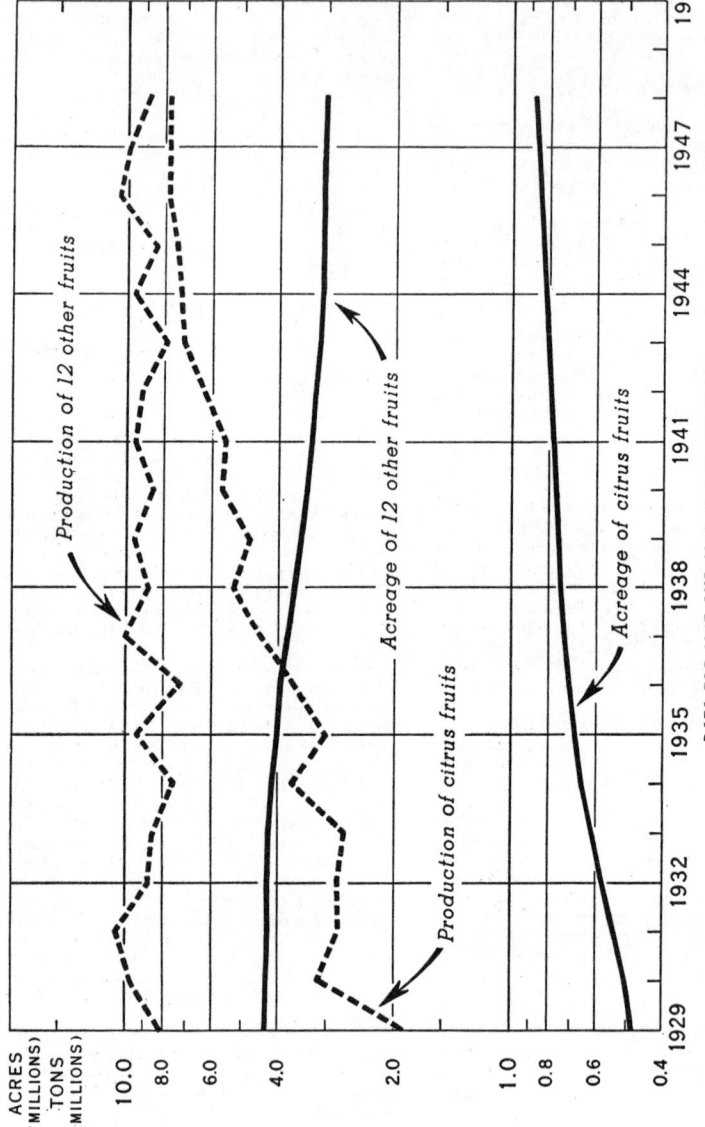

FIGURE 28.—ACREAGE AND PRODUCTION OF CITRUS FRUITS AND OF 12 OTHER MAJOR FRUITS, UNITED STATES, 1929–48.

DATA FOR 1947 AND 1948 ARE PRELIMINARY.

Acreage of citrus fruit has steadily increased since 1930, whereas the acreage of deciduous fruits has declined. The increase in production of all fruits is due almost entirely to the rapid increase in production of citrus fruits; these fruits account for almost one-half of total production on a tonnage basis. This is a ratio chart showing equal percentage changes.

stimulated heavy feeding as well as the breeding of more animals for 1949 production.

In addition to the changes in the supply of feed there have been notable changes in quality, resulting in the feeding of rations that are more balanced with respect to protein and other nutritive elements. The gradual increase in the protein content of the hay supply has been mentioned (fig. 19, p. 34). The greater supply per animal unit of oilseed meals and other high-protein feeds during the war also helped to balance the feeding rations and to push livestock production upward.

PRODUCTION PER UNIT OF BREEDING STOCK

Changes that have taken place since 1919 in production per unit of breeding livestock are shown for all cattle and for hogs in figure 29, page 48. Figure 5 showed changes in the total production of livestock for human use per animal unit of breeding livestock. This series of records of production per animal unit of breeding livestock are similar to the series that show changes in yields per acre of different crops and production per acre of all crops. They measure the combined effects of all the forces that are back of the changes in livestock production *per unit of breeding stock*. Similarly, the index of production of all crops per acre, measures the combined effects of all the forces that operate to change the total crop production *per acre*.

Changes in livestock production per animal unit of breeding livestock are influenced by weather, chiefly because feed supplies are affected by favorable or unfavorable growing conditions. The bad drought years of the 1930's are reflected in reduced production per unit of breeding animals. But the downward trends shown for those years in figure 29 probably also reflect less attention to livestock in other ways because of the depression and the low prices. On the other hand, the years that combine large feed supplies and favorable prices for livestock soon show the increased production that results both from more liberal feeding and from better care in other ways. For example, in cattle production a larger calf crop, a reduction in death loss, and prevention of disease, all combine with better feeding and other care to produce more meat and milk per cow in the breeding herd.

Increase in output per unit of feed is one measure of increased efficiency in livestock production. There is evidence that significant gains have been made in this direction in all classes of livestock, but the data available on a national basis are not sufficiently refined to allow these changes to be traced. Information obtained in the Corn Belt over a period of years on feed consumed by hogs indicates a reduction of 10 to 15 percent in the quantity of feed used per 100 pounds of pork from the decade of the 1920's to the decade of the 1930's.[5] The number of pigs saved, the prevention of disease, the improvement of breeds, and the feeding of balanced rations all make for higher efficiency in the use of feed.

If similar data were available for tracing changes in feed consumed per 100 eggs produced it seems probable that even greater reductions in feed per unit of product would be shown. Even the over-all national estimates indicate a reduction of about 12 percent in concentrates consumed per 100 eggs produced from the period 1920-24 to 1937-41.[6]

Figure 30, page 49, shows the trend of egg production per layer for the average number of layers on farms. This series shows a sharp upward climb, especially within recent years. The trends in egg production reflect the noteworthy improvements that have been made in the poultry enterprise. In many parts of the country, and especially in the Midwest, it has been transformed from a sideline to an important phase of the farm business.

Figure 31, page 50, shows the changes in milk production per cow over nearly four decades. There was a sharp increase in the 1920's, and then a drop from 1929 to 1934, reflecting the drought and depression then prevailing. More cows were milked during the depression than in the 1920's, so a larger proportion of the cows milked were of beef or mixed breeding, and the feed supply was reduced drastically by drought. Milk production per cow did not return to 1929 levels until 1938. It was maintained at high levels during the war and has reached successive record peaks since that time. Preliminary figures indicate an all-time record of 5,036 pounds per cow in 1948.

Production per cow seems high in relation to previous years but there is still room for considerable improvement. The average milk production per cow, for cows on which full-year records were kept by dairy herd improvement associations in 1945, was 8,592 pounds compared with a national average of 4,797 pounds in that year. In other words, the cows in dairy herd improvement associations produced about 80 percent more milk per cow than the national average. These two estimates of production per cow are not strictly comparable, but the wide difference between the two figures indicates the potentialities of greater production per animal.

More feed per cow, and better balanced rations, would be the two most influential factors in achieving a higher national average production per cow.

---

[5] ATKINSON, L. JAY, AND KLEIN, JOHN W. FEED CONSUMPTION AND MARKETING WEIGHT OF HOGS. U. S. Dept. Agr. Tech. Bul. 894, 28 pp., illus. 1945. See pp. 19–21.

[6] JENNINGS, R. D. FEED CONSUMPTION BY LIVESTOCK, 1910–41—RELATIONS BETWEEN FEED, LIVESTOCK, AND FOOD AT THE NATIONAL LEVEL. U. S. Dept. Agr. Cir. 670, 57 pp. 1943. See p. 28.

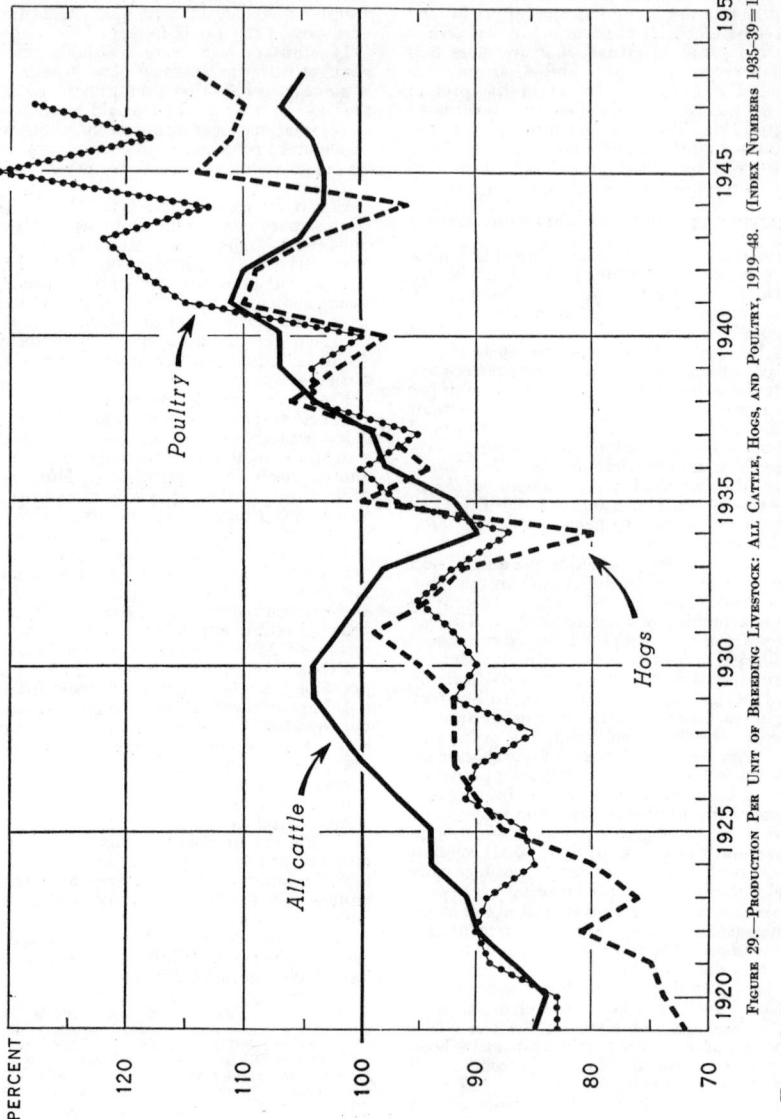

FIGURE 29.—PRODUCTION PER UNIT OF BREEDING LIVESTOCK: ALL CATTLE, HOGS, AND POULTRY, 1919–48. (INDEX NUMBERS 1935–39=100.) Hog production per unit of breeding stock has moved sharply upward since 1919. There has also been an upward trend in cattle production per unit.

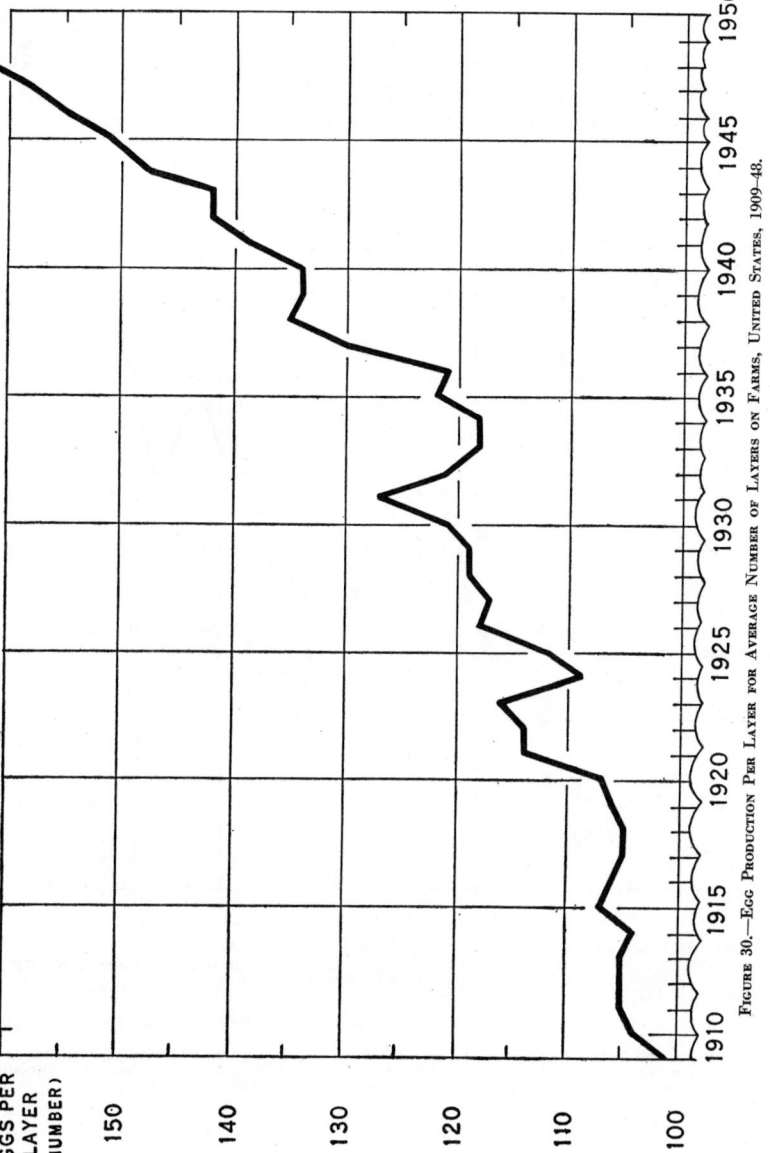

FIGURE 30.—Egg Production Per Layer for Average Number of Layers on Farms, United States, 1909-48.

Egg production per layer made an upward climb of 50 percent from 1909 to 1945. In the spring of 1949 there was still no indication of leveling off.

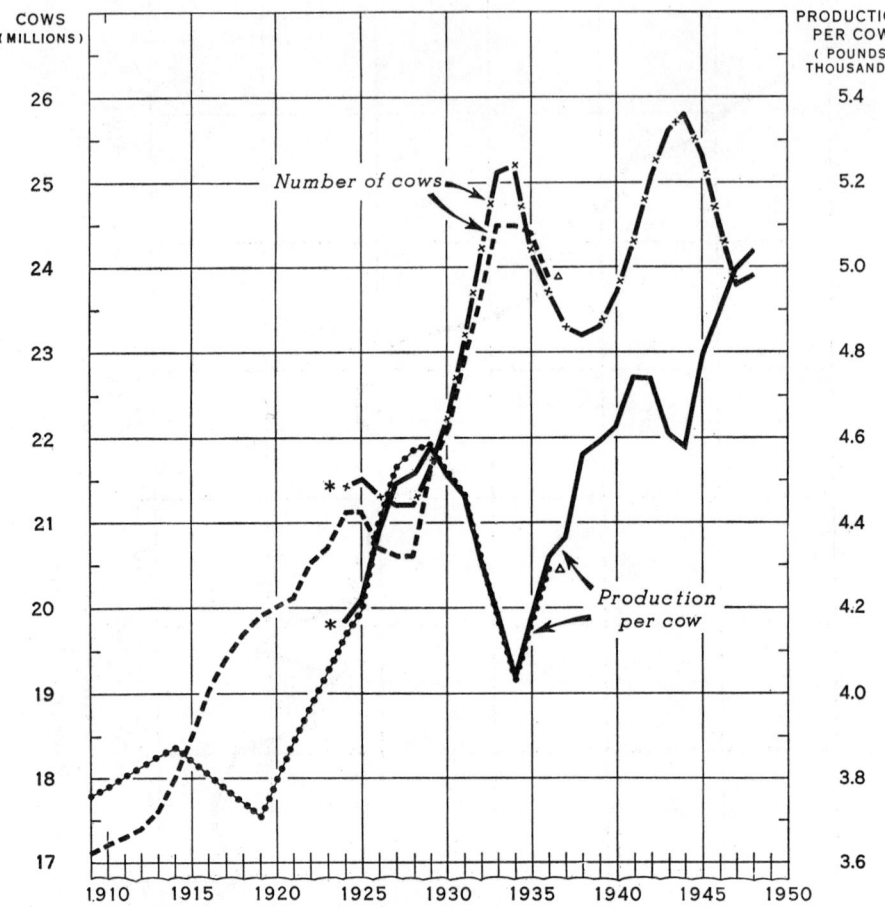

* FROM BAE REPORTS "PRODUCTION, DISPOSITION, AND INCOME FROM MILK"
△ COMPUTED FROM TABLES A-95 AND A-96 WPA REPORT, "TRENDS IN SIZE AND PRODUCTION OF THE AGGREGATE FARM ENTERPRISE, 1909-36"
DATA FOR 1948 ARE PRELIMINARY

FIGURE 31.—MILK: NUMBER OF COWS MILKED AND PRODUCTION PER COW MILKED ON FARMS, UNITED STATES, 1909–48.

A definite upward trend in milk production per cow for the period since 1909 was interrupted briefly during World War I, and then for a longer time in the drought and depression years of the 1930's (when many cows usually kept for beef were milked). An all-time peak in production per cow was reached in 1948.

Because such a large proportion of the feed used by a dairy cow is required for maintenance, underfeeding means that most of the feed is used for body maintenance, and a relatively small part of it for milk production. In the newer dairy areas, and where dairy production is only a sideline, underfeeding is rather common.

It is recognized, of course, that there are economic limits to heavier feeding, and that these are reached before the physical limits are approached. The economic limits will vary with the price of the product in relation to the cost of feed, and in relation to other expenses. This is true in all classes of livestock, but many farmers feed their cows at levels far below the economic limits, even when milk is high in price in relation to feed. To increase the feed supply per animal, for dairy cows as well as for other classes of livestock, would increase both output per head and net income on the majority of farms.

Other factors also will tend to increase production per animal in future years. New developments in cross-breeding show considerable promise. In dairy cattle, artificial insemination makes it possible to develop the higher milk-producing strains more rapidly than previously. On many farms the herd bull will disappear, and the feed will be fed to milk cows. Improvements in hog breeding, already under way, will increase efficiency in use of feed and produce a carcass of higher quality. More progress will be made in control of disease for all classes of livestock.

PROSPECTIVE DEVELOPMENTS

More progress is called for in improving shelter and equipment, in order to reduce labor and capital requirements in caring for livestock. Further improvements are badly needed in the quantity and quality of the feed supply and in reducing the cost of making the feed available to livestock. As farmers shift toward more soil-conserving systems of farming, more hay and pasture will be available, and that means more roughage-consuming livestock. But improved techniques are needed to produce the roughage in ways that will result in more livestock and livestock products at lower cost per unit of product.

As the introduction of mechanical power gradually makes further progress in the South, the numbers of productive livestock are likely to increase, because of the feed that will be released and because systems of farming are likely to be developed that include more hay, grain, and pasture—crops that can be handled by mechanical equipment. Cattle numbers have increased relatively more in the humid areas of the country than on the ranges, of late, because of the shift to tractor power. This proportionately greater increase will probably continue. On the ranges, expansion of livestock production is dependent upon improvements that will increase carrying capacity, and upon developments that will make it possible to produce more winter feed.

CHANGES IN FARM SIZES AND OWNERSHIP

Factors responsible for a large part of the increases in production also have had considerable influence on changes in the number and sizes of farms (table 6 and fig. 32, p. 52). A part of the change in sizes since 1920 results from factors related to development of new arable land in the West and abandonment of land in the East; and to the very considerable growth in part-time farming, and establishment of rural homes by those engaged in nonfarm work.

The total number of farms counted in the census of agriculture decreased 9 percent from 1920 to 1945. On the other hand, the "land in farms" increased 19 percent. The latter change occurred mostly in the 17 Western States. In fact, the land in farms decreased in most of the Eastern States during this period.

TABLE 6.—*Number of farms by size groups in the United States, census years 1920, 1930, 1940, and 1945*

Census year	Acreage size group							
	Total	Under 10	10–19	20–99	100–259	260–499	500–999	1,000 and over
	Thousands	Thousands	Thousands	Thousands	Thousands	Thousands	Thousands	Thousands
1920	6,448	289	508	2,978	1,980	476	150	67
1930	6,289	359	560	2,815	1,863	451	160	81
1940	6,097	506	559	2,512	1,796	459	164	101
1945	5,859	594	526	2,286	1,693	473	174	113

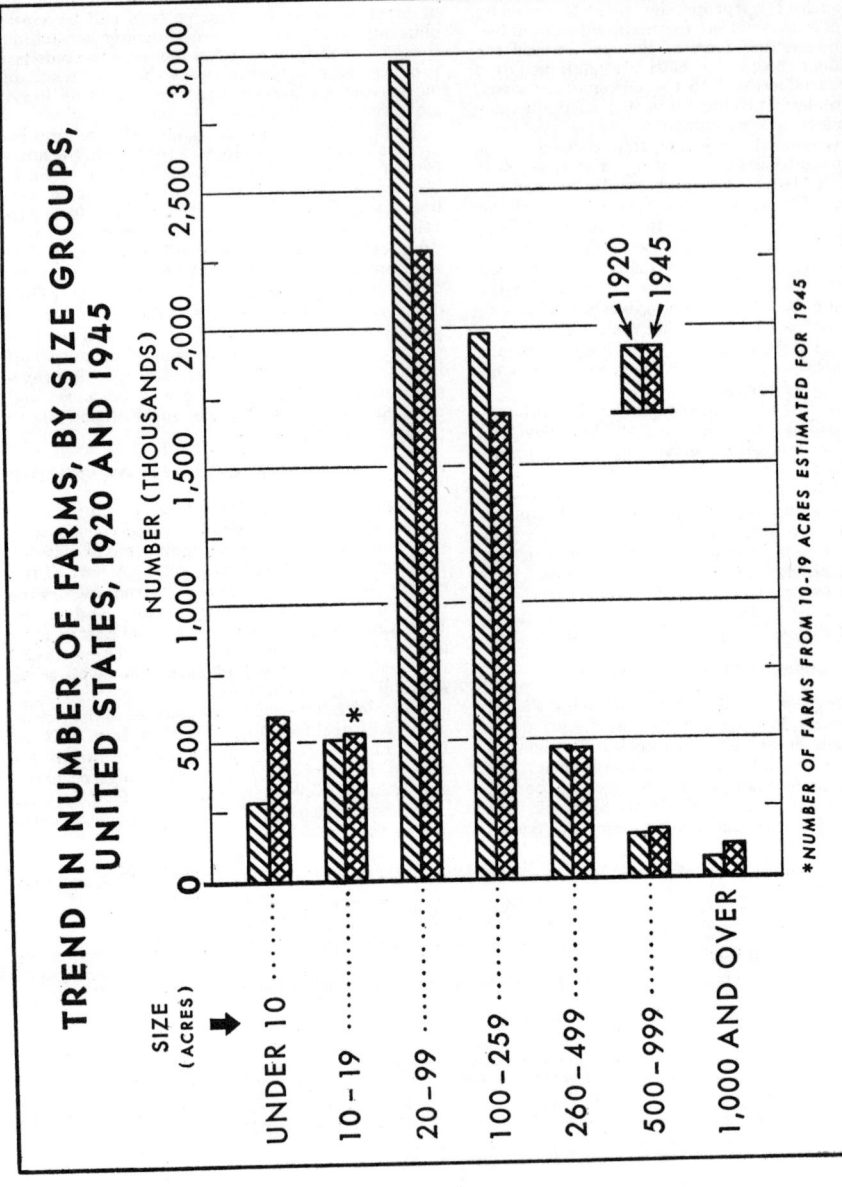

FIGURE 32.—IN THE QUARTER-CENTURY FROM 1920 TO 1945 THE NUMBER OF FARMS UNDER 10 ACRES ENUMERATED BY THE CENSUS MORE THAN DOUBLED. THE NUMBER OF FARMS IN THE SIZE GROUPS FROM 20 TO 259 ACRES DECREASED SUBSTANTIALLY; AND THERE WAS A LARGE PERCENTAGE INCREASE IN THE FARMS OF MORE THAN 1,000 ACRES, WHICH WAS LARGELY ACCOUNTED FOR BY WHEAT FARMS AND LIVESTOCK RANCHES IN THE WESTERN STATES.

MORE PART-TIME FARMS AND RURAL HOMES

In the quarter-century from 1920 to 1945 there was a 106-percent increase in the number of extremely small units that are counted as farms by the census—those under 10 acres (table 6 and fig. 32). Farms of that size are mostly part-time farms, rural homes, and retirement units. Very few are considered as actual farms in the localities where they are found. But they are counted by the census as farms because they have 3 acres or more, or have value of products of $250 or more. The number of farms from 10 to 19 acres increased slightly from 1920 to 1945; these also are frequently part-time farms.

Table 7 contains estimates by the Bureau of Agricultural Economics of the number of part-time and residential units that were counted as farms in the censuses of 1930, 1940, and 1945. This table also shows the number of farms, for those years that may be regarded as "farming units"—farms that are operated primarily as a source of income or to provide a living for the farm family rather than being primarily a place to live. There were 442,000 more part-time and residential farms in 1945 than in 1930, and 871,000 fewer farming units. In 1945 only about 4,300,000 farms could be classed as farms that were primarily a source of income or living for the farm family, rather than being primarily a place to live.

TABLE 7.—*Changes in number of census farms, farming units, and part-time and nominal units, 1930, 1940, and 1945* [1]

Kind of unit	1930	1940	1945
	Thousands	Thousands	Thousands
All census farms	6,289	6,097	5,859
Farming units	5,141	4,752	4,270
Part-time and nominal units	1,148	1,345	1,589

[1] Number of "farming units" and "part-time and nominal units" for 1930 and 1940 estimated from census data; 1945 numbers derived from "Special Report 1945 Sample Census of Agriculture," table 29.

FEWER SMALL FARMING UNITS

In contrast to the large increase in numbers of the extremely small part-time and residential farms, from 1920 to 1945, there was a 23-percent decrease in what might actually be called "small farms," those with 20 to 99 acres. There were 692,000 fewer farms in this group in 1945 than in 1920. There were 14 percent fewer farms in the size group 100 to 259 acres; this group includes the traditional 160-acre homestead size. But the group from 260 to 499 acres nearly held its own; it showed only a 1-percent decrease in number of farms from 1920 to 1945.

TREND TOWARD LARGER FARMS

At the upper end of the size-of-farm scale there was an increase in the number of farms during this period. The group from 500 to 999 acres showed an increase of 16 percent, and those of 1,000 acres and over increased 69 percent. Although the group of farms of 1,000 acres and over was two-thirds larger than it was in 1920, that group still contained less than 2 percent of the total number of farms, in 1945. But operators of farms of that size controlled about 40 percent of the total land in farms. This seems like a rapid trend toward concentration of land holdings until the data are analyzed more closely. About 87 percent of the number of farms of 1,000 acres or over were found in the 17 Western States. This means that the increase took place mostly in the ranching and dry-land wheat area where 1,000 acres is not a large-scale farm. But census data and other available information indicate that there has actually been some increase in the number of farms that might be termed large-scale farms outside of the grain and ranching areas of the Western States.

More noteworthy than the growth in large-scale farming, was the shift to larger family farms within all the size groups of 100 acres and over. It was made possible and has been accelerated by technological changes, especially by adoption of mechanical power and complementary equipment.

CLASSIFICATION OF FARMS IN 1945

Tables 8 and 9 and figure 33, page 55, provide a summary picture of the distribution of farms by economic classes in 1945. The classification of the farming units is chiefly on the basis of value of products as a measure of size. No comparable figures are available for previous census periods, which means that it is not possible to trace changes in these classes over a period of years. But the 1945 data indicate that farming in this country is still preponderantly a family enterprise. Although the large-scale farm group included 26 percent of the farm acreage and 22 percent of the value of production, the three family farm groups had more than 60 percent of the acreage and produced more than 70 percent of the value of farm products.

TABLE 8.—*Percentage of farms, population, acreage, and value of farm products by economic class, United States, 1945*

Economic class [1]	Number of farms	Farm population	Farm acreage	Gross value of farm production
	Percent	Percent	Percent	Percent
Farming units:				
Large-scale farms	1.7	3.7	25.8	21.9
Commercial-family farms:				
Large	7.0	8.5	18.3	23.5
Medium	20.0	21.3	24.1	30.0
Small	28.4	28.5	18.1	17.1
Small-scale farms	15.8	14.0	5.8	4.2
Other units:				
Part-time units	10.3	10.9	2.3	1.9
Nominal units	16.8	13.1	5.6	1.4
All farms	100.0	100.0	100.0	100.0

[1] Special Report 1945 Sample Census of Agriculture, table 29. Economic class is defined in terms of the *total value of products* sold and used by the farm household modified by specified secondary criteria: Large-scale farms, $20,000 and over; large family farms, $8,000 to $19,999; medium family farms, $3,000 to $7,999; small family farms, $1,200 to $2,999; small-scale farms, $500 to $1,200; part-time units, $250 to $1,200 with operator working off farm 100 days or more; nominal units, less than $500 with some adjustments for work off farm and abnormal relative values of farm products and land and buildings.

Perhaps the most difficult farm problems are found on the nearly 1,000,000 small-scale farms that had less than 6 percent of the total acreage and produced only 4 percent of the farm products. The annual value of products on these farms is from $500 to $1,200. Although the farms on which the operator worked off the farm 100 days or more are not included in this group, we do not know how many of these farm families had other sources of income. It is safe to assume, however, that a large group of them had extremely low incomes available for living even in the relatively prosperous year of 1944. These small-scale farms tend to be concentrated in such areas as the Southern Appalachians and the cut-over parts of the Lake States.

Looking forward, some of the same forces are likely to continue to influence changes in the number and sizes of farms as have operated over the last quarter-century. We might expect a further large increase in the number of part-time farms and rural homes. If nonfarm employment is available there might be a gradual decrease in the number of small-scale farms. The full-time family-operated farms are likely to be fewer and larger. And there might be some further increase in the number of large-scale farms, but they will still constitute a relatively small percentage of the total number of farms.

TABLE 9.—*Number and important characteristics of farms by economic class, United States, 1945*

Economic class [1]	Number of farms	Average per farm				
		Gross value of product	All land	Harvested crop land	Value, land and buildings	Value, implements and machinery
	(1,000)	Dollars	Acres	Acres	Dollars	Dollars
Farming units:						
Large-scale farms	102.1	39,203	2,905	384	78,422	6,452
Commercial-family farms:						
Large	408.9	10,484	514	193	26,067	3,021
Medium	1,173.0	4,658	236	104	11,135	1,616
Small	1,661.9	1,874	125	46	5,117	595
Small-scale farms	923.5	825	72	22	2,305	204
Other units:						
Part-time units	602.2	574	43	10	2,585	209
Nominal units	987.3	264	65	11	3,583	176

[1] Special Report 1945 Sample Census of Agriculture, table 29. Economic class is defined in terms of the *total value of products* sold and used by the farm household modified by specified secondary criteria: Large-scale farms, $20,000 and over; large family farms, $8,000 to $19,999; medium family farms, $3,000 to $7,999; small family farms, $1,200 to $2,999; small-scale farms, $500 to $1,200; part-time units, $250 to $1,200 with operator working off farm 100 days or more; nominal units, less than $500 with some adjustments for work off farm and abnormal relative values of farm products and land and buildings.

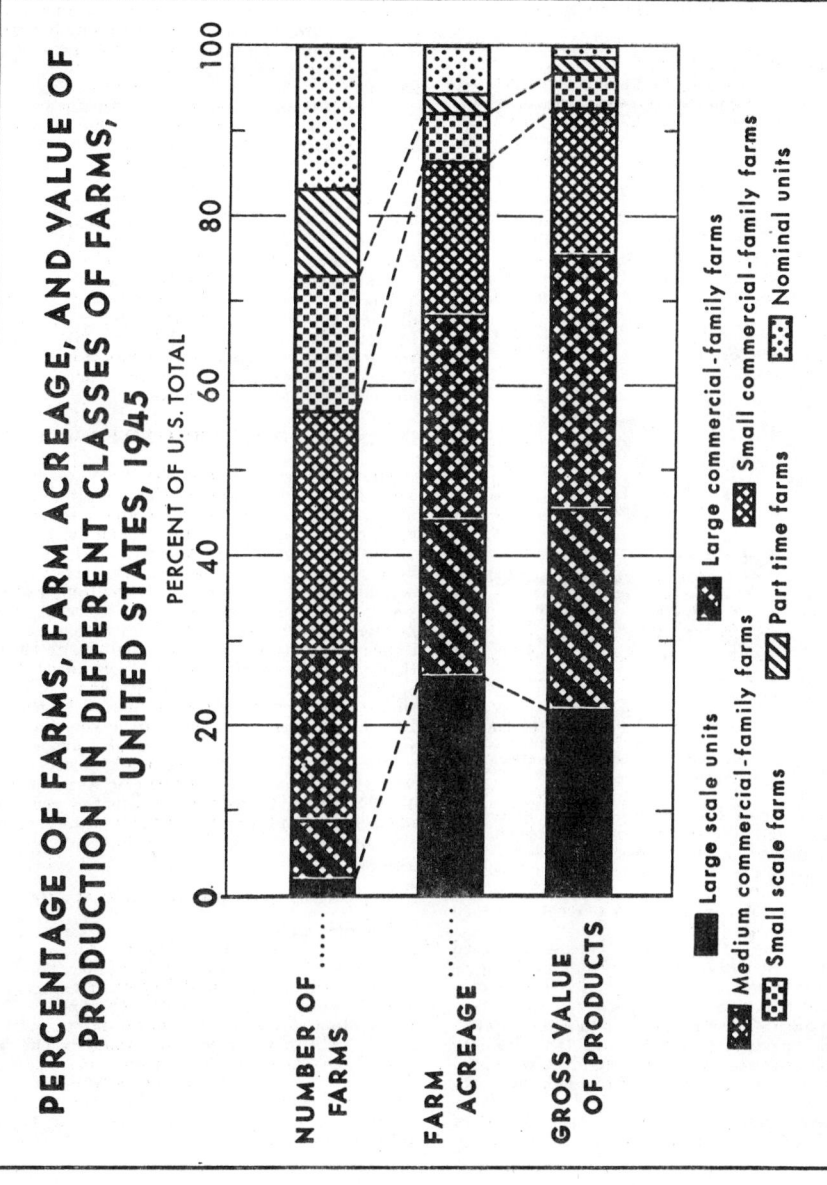

FIGURE 33.—IN 1945 MORE THAN ONE-FOURTH OF THE FARMS COUNTED BY THE CENSUS WERE CLASSIFIED AS PART-TIME AND NOMINAL UNITS, BUT THEY HAD ONLY 8 PERCENT OF THE FARM ACREAGE AND PRODUCED ONLY 3 PERCENT OF THE GROSS VALUE OF PRODUCTS. ON THE OTHER HAND, LESS THAN 2 PERCENT OF THE FARMS WERE CLASSIFIED AS LARGE-SCALE, BUT THEY HAD MORE THAN ONE-FOURTH OF THE FARM ACREAGE AND PRODUCED MORE THAN ONE-FIFTH OF THE FARM PRODUCTS.

RECENT TRENDS IN FARM OWNERSHIP

There have been noteworthy changes in farm tenure during the last quarter-century. Census returns for 1945 indicate that about 32 percent of all the farms counted by the census were operated by tenants, as contrasted with 38 percent in 1920. Tenancy increased in the decade following 1920, and 42 percent of the farms were operated by tenants in 1930. But by 1940, the percentage of tenancy was about back to 1920 levels. A considerable decrease between 1940 and 1945 resulted in the lowest level of tenancy since before 1900. Information from a later survey by the Bureau of Agricultural Economics indicates that only 27 percent of the farms were operated by tenants in 1948.

TABLE 10.—*Number of farms by tenure of operator in the United States, census years 1920, 1930, 1940, and 1945*

Census year	All operators	Tenure of operator [1]			
		Full owners	Part owners	Share croppers [2]	Tenants other than croppers
		Thousands	Thousands	Thousands	Thousands
1920	6,448	3,366	559	561	1,894
1930	6,289	2,912	657	776	1,888
1940	6,097	3,084	615	541	1,820
1945	5,859	3,301	661	447	1,412

[1] Excludes managers.
[2] Sharecroppers are concentrated in the cotton and tobacco areas of the Southern States. The landlord usually furnishes all the power and equipment, and the cropper provides the labor. Cropper operations are usually closely supervised.

The number of full owners actually increased 7 percent from 1940 to 1945, at a time when the total number of farm operators decreased 4 percent. A large part of the increase in farm ownership is accounted for by the greater number of farms under 10 acres, about 75 percent of which are owner-operated.

The number of part-owner farms increased about 18 percent from 1920 to 1945, and the acres of land they operated by 112 percent. The greater number of part-owner farms helps to explain how so many farms have increased in acreage. Farmers who owned some land have rented adjoining farms or separate tracts that could be combined with their own land for operation as a more efficient unit.

Owner-operatorship of family farms is one of the goals of agricultural policy. The tenure figures for 1945 and 1948 indicate considerable recent progress toward that goal. Data on mortgage debt also indicate that farmers have greatly increased their equity in the land they own. Only 29 percent of the farms in this country had mortgages in 1945, compared with 39 percent in 1940. The total farm-mortgage debt shrank from 6.6 billion dollars in 1940 to a low of 4.7 billion dollars in 1946. But the downward trend was reversed from 1946 to 1947. Mortgage debt has continued to increase since that time, and on January 1, 1949, it was 9 percent above the 1946 low point. There have been large increases in several States of the East, South, and West, that were partly offset by continued reductions in the Midwest.

The blind spot in the mortgage situation is the distribution of mortgage debt among individual farmers. If a large part of it is on farms where young men have made commitments at high prices financial trouble spots are likely to develop with any downturn in farm incomes.

INCENTIVES FOR INCREASED PRODUCTION

That farm production in the 1930's was held in check by drought and depression, despite the technical progress in mechanization and in other lines, has been emphasized. War needs and the incentive of higher incomes broke this dam and released a flood of increased production. But this was not accomplished by one blast at the beginning of the war. It was a fairly gradual process. Both farmers and agricultural workers were too conditioned by the experience with surpluses to believe that the market really would absorb all that the agricultural plant might produce under stimulation. This skepticism was supported by the sag in prices of some products during the year that followed the outbreak of the war in Europe.

But gradually the war demands emerged, and with the passage of the Lend-Lease Act on March 11, 1941, they gathered a momentum that remained unslackened for the duration of the war. Demand was further accelerated in the first two postwar years.

INITIATION OF WARTIME PRICE SUPPORTS

The first public pronouncement of the need for increasing the production of food was made by the Secretary of Agriculture on December 26, 1940, when he urged the desirability of breeding more sows for spring farrow in 1941. On April 3, 1941, price supports were announced for dairy products, hogs, chickens, and eggs, that would be effective until June 30, 1943. This announcement assured farmers of a market for expanded production of these products. But production controls were maintained on wheat, cotton, tobacco, and corn, for the years 1941 and 1942. They were removed from corn in January 1943, from wheat in February 1943, and from cotton in July 1943.

Part of the reason for maintaining production controls in the early war years was to obtain a shift in the direction of meat, eggs, milk, and oil crops—the products most urgently needed in the first part of the war. But a more effective means was needed for achieving the most desirable combination of farm products. The program of formulating and announcing production goals, combined with support prices and with production payments that were geared to the relative urgency of need for different products, was developed to serve this need.

DEVELOPMENT OF PRODUCTION GOALS

The first production-goals program was worked out in the summer and fall of 1941. It outlined the food needs for 1942 and stated the production objective for each product. These goals were revised in January 1942, following the attack on Pearl Harbor. Goals were successively developed for each of the succeeding years.

Production goals were one step in the process of arriving at a balanced production program. Before goals were determined for any product the prospective needs were analyzed in relation to the resources and the facilities available for its production. Then each product was considered in relation to all the other farm products that were needed in a balanced production program. Studies of production capacity for individual products, and for all products combined into a production program geared to prospective needs, were made in each State to provide a production guide for the goals program. The program as finally developed pointed the direction and indicated the distance that should be traveled to achieve a balanced production.

PROGRAMS FOR ACHIEVEMENT OF GOALS

Support prices and production payments, when balanced in relation to the desired production of each product, furnished most of the driving force that was needed to achieve the objectives outlined in the goals program. But education and information concerning war needs also played an active part, and patriotism helped as the war progressed. Many farmers grew soybeans, peanuts, flax, dry beans, and other strategic war crops because they knew the need was urgent, even though they might have obtained somewhat larger returns by growing other products.

So far as possible, however, support prices and production payments were intended to make the most urgently needed products the most profitable to the producers. Efforts in this direction were somewhat limited by minimum loan provisions for some crops not so urgently needed, and by price ceilings on others; but in general a pattern of support prices and payments was eventually achieved that was reasonably well balanced in relation to the war needs for each product.

With production goals to point the direction and to indicate the distance to be traveled; with support prices, production payments, and educational persuasion; with patriotism and family participation in the war as the fighting continued, as further incentives to attainment, a level and pattern of production was finally achieved that was fairly well proportioned to war needs. Some of the demands for food were not satisfied. They could have been met only by devoting more equipment, and materials to agriculture at the expense of other sectors of the war program. But the most urgent needs were satisfied, and the pattern of production was shifted in the direction of products with the highest war priorities. It seems doubtful that the large changes in production shown in figure 2 could have been accomplished without a program that emphasized the need for those changes, and that supplied incentives for obtaining them. Greater shifts from livestock products to direct food crops could have been made if the food requirements that were developed had called for more sacrifice in quality in order to provide food for more people. And if the need had been so urgent as to have forced a larger allocation of resources to agriculture it would have been possible to obtain much greater increases in production.

So far as materials were concerned, farmers felt the greatest pinch in new farm machinery, especially for the production year 1943, when only 23 percent of the 1940 volume of steel used for farm machines was originally allocated for that purpose. This allocation was increased later, but new machinery was unobtainable for most farmers in 1943. More farm machinery could have been substituted for labor, and the process of mechanization would have advanced further by the end of the war. This in turn would have facilitated agriculture's adjustment to peacetime conditions.

Shortages of some other materials developed early in the war. Fencing, building materials, containers, and other items were scarce; but the minimum needs were met by careful distribution of available supplies. Fortunately, the supply of insecticides and of commercial fertilizer was fairly ample, although more would have been used if it had been available.

Scarcity of labor constituted the worst obstacle to production in some of the seasonal cash-crop areas. The Office of Labor and the State Extension Services assisted in bringing in outside labor, and in recruiting and training local labor from previously untapped sources. Farm families often worked long days to get the essential jobs done.

POSTWAR DEVELOPMENTS

The end of the war relieved some of the scarcities of labor and materials. Farm employment averaged 92 percent of 1935–39 in 1946, 93 percent in 1947, and 92 percent in 1948, compared with 90 percent in 1945. New machinery was not available in sufficient volume to supply all farmers with all the machines they would like to buy until the spring of 1949, when most scarcities disappeared.

Production goals and support-price programs were in effect for the years 1946–48, but the price-ceiling structure lapsed temporarily in July 1946 and, after a short period of reinstatement, was removed from nearly all products in the fall of 1946. Most farm products sold at prices above support levels until the summer of 1948, although potatoes, eggs, and some other commodities required Government support at different times.

Prices received by farmers were 204 percent of their 1935–39 average in June 1946. In October 1946, after most of the price ceilings were removed, they were 255 percent of prewar. In January 1948 they had risen to 287 percent. In January 1949 they were 250 percent of 1935–39, and in April 1949 they were 243 percent.

It is evident that price incentives were even better in the first two postwar years than during the war. On the expense side, however, farm costs have also risen, but not so rapidly as prices received. Prices paid for goods and services used in farm production (not including farm wages) were 150 percent of 1935–39 in June 1946. They had risen to 163 percent in December 1946, and to 199 percent in December 1947. For the entire year of 1948 they averaged 201 percent of prewar, and in April 1949 they were 192 percent. Farm wage rates in the prewar years, 1935–39, were only about two-thirds of the level prevailing in the 1920's. But they rose very rapidly during the war, and in June 1946 they were 321 percent of 1935–39. They averaged 346 percent of prewar in 1947, and 367 percent in 1948.

With cost rates lagging behind the rise in farm prices, and with a much larger volume of output of marketable products, net farm incomes have increased a great deal. Figure 34, page 59, summarizes the gross and net farm-income results to the farmers of the country of their production job during the war and early postwar years. The realized net income of farm operators for the war years 1942–44 averaged 240 percent of the 1935–39 level. In 1946 it was 324 percent of the average for those years; and in 1947 it was 386 percent—nearly four times the net income of prewar years. The rise in net incomes from 1946 to 1947 is almost entirely attributable to changes in prices because the volume of production was slightly lower in 1947 than in 1946. In 1948 the net income was somewhat lower because of the decline in prices for farm products and rising rates of costs, but it still averaged 364 percent of 1935–39.

The operator's net farm income on family-operated farms for different types and locations is shown in figure 35, page 60. This chart traces the extremely low net incomes that prevailed during the early 1930's, the slow recovery during the latter part of that decade, and the rapid rise in the war and early postwar years. The greatly increased production per farm in recent years has meant relatively lower expenses per unit of product; and with more products to sell the net incomes rose faster than did the prices received for those products. If farmers should encounter several consecutive years of lower prices with cost rates remaining at or near present levels, their margin between expenses and gross income would narrow; and net incomes would be reduced faster than the drop in farm prices unless efficiency could be increased to reduce costs per unit of product and thus to offset the lag in cost rates.

IMPLICATIONS OF RECENT AND PROSPECTIVE CHANGES

The forces that shaped the course of agricultural production in the interwar, the war, and the early postwar years, have been analyzed briefly in these pages. Their effects on production are evident in the record-breaking volume of recent years. Most of these forces still have unexpended power. They will continue to influence production in the years beyond the transition from war to peace. New forces, expected and unexpected, will be set in motion. Always farmers will need to adapt their operations to the rapidly changing conditions.

PROSPECTIVE CHANGES

Assuming that a stable peace can be established, and then looking forward beyond the transition years to the time when farming will be adjusted to peacetime conditions, some changes seem fairly certain. They will result from the operation of the forces now under way, and of those that are on the horizon. The changes that seem most likely to occur are summarized as follows:

1. A continuation of the shift to mechanical power until it has largely supplanted animal power is to be expected. The smaller tractors that are more suitable for small farms and rolling land will accelerate this shift in the South, and in other areas that have small farms.

2. Further adaptation of machines for use with mechanical power is certain. Each phase of agricultural production will become more mechanized,

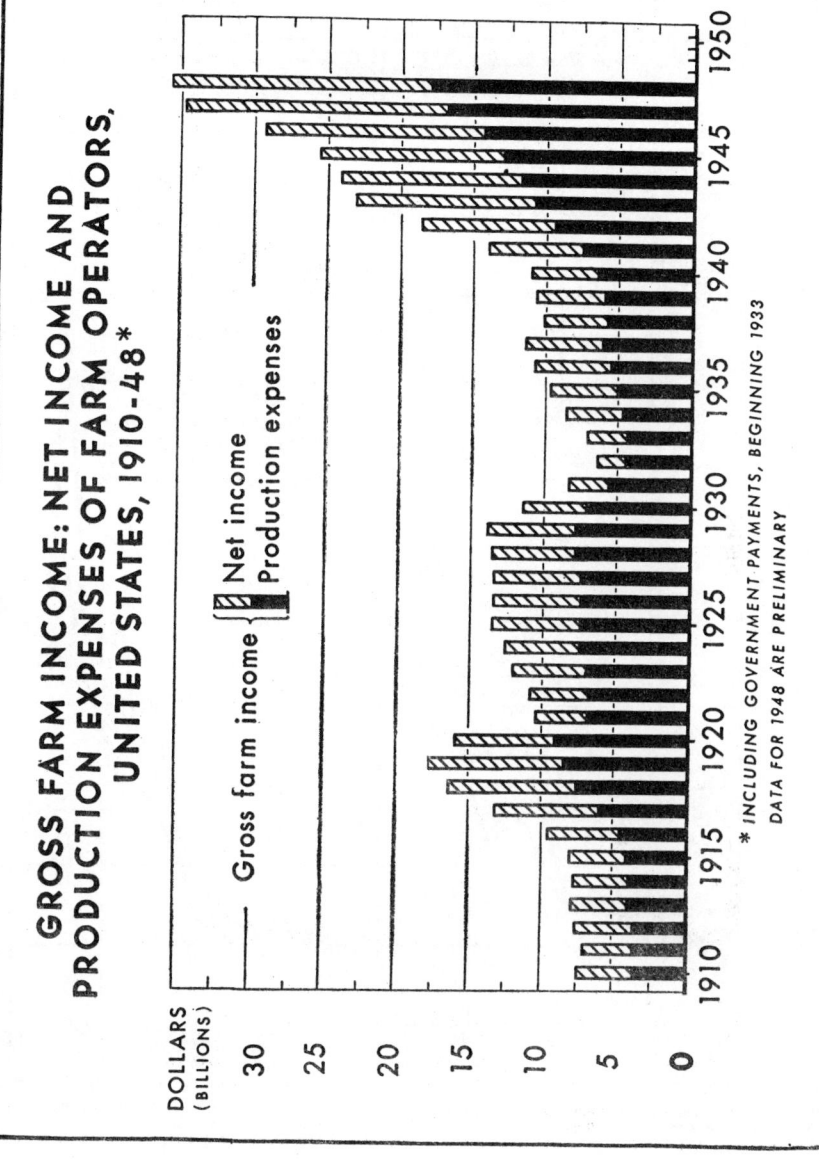

FIGURE 34.—PRODUCTION EXPENSES OF FARM OPERATORS HAVE INCREASED SINCE 1940 BUT GROSS FARM INCOME ROSE MORE RAPIDLY THAN EXPENSES UNTIL 1948.

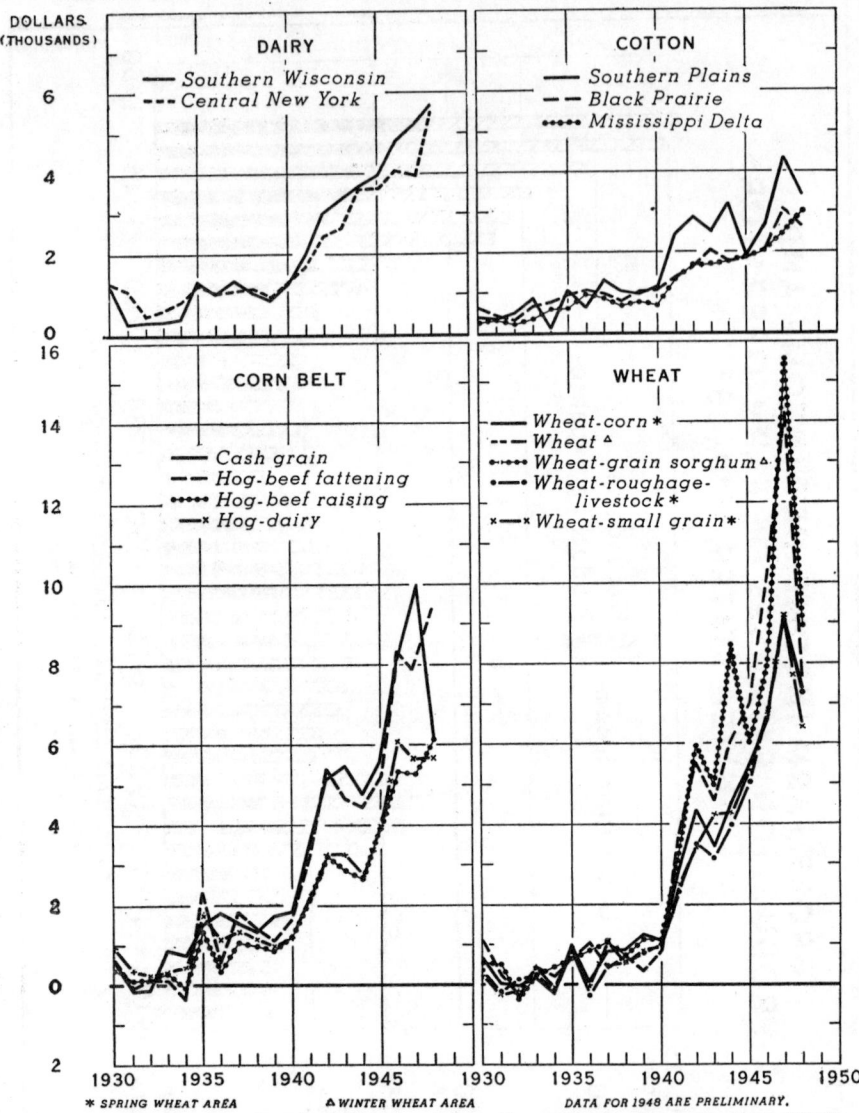

FIGURE 35.—FARM EARNINGS: OPERATOR'S NET FARM INCOME, FAMILY-OPERATED FARMS, BY TYPE, 1930–48.

Annual earnings of farm families operating all types of farms seem to follow the same general pattern. The extremely low net incomes of the drought and depression years 1931–36 were followed by some recovery in the late 1930's. The sharp rise in the war and early postwar years resulted from higher prices combined with increased production per farm.

and more fully adapted to mechanical-power techniques. Eventually the same type of stability may be achieved with mechanical power as was attained with machinery adapted for animal power, previous to World War I. But this process will require some time for full development.

Haying equipment will be adapted to the special conditions of each producing area. Mechanical cotton pickers and strippers will be adopted, gradually at first, and then more rapidly as changes are made in them for more effective use in areas of rolling land and small farms. Some progress may be expected in mechanizing the production of such traditionally hand-labor crops as sweetpotatoes and even tobacco.

3. Use of lime and commercial fertilizer will rise above the high 1947 levels. How rapidly this increase will come depends partly on the kind of educational and conservation programs that are developed, and partly on the level of farm incomes. But many farmers have now learned the value of lime and fertilizer, and they are not likely to reduce their purchases except under conditions of severe depression.

4. Along with the use of more lime and fertilizers will come more rapid adoption of other conservation practices, such as using winter cover crops, grass and legume crops in the rotations, and following contour farming, strip cropping, and other practices designed to control erosion.

5. Further progress will be made in varietal improvements. For example, Lincoln soybeans and Clinton oats are now being adopted. Suitable corn hybrids are being developed and will be adopted in the Southern States. The effects of improved varieties of grain sorghums will become more pronounced.

6. Progress will be made by farmers in combining the use of improved varieties, lime and fertilizer, and conservation and other practices, in effective crop rotations and systems of farming that will result in much higher production because the combined effects of these improvements will be greater than if they are adopted as single practices. In North Carolina, for example, a number of corn experiments combining high nitrogen fertilization, hybrid seed, and other improved practices, resulted in yields of more than 80 bushels per acre compared with usual yields of 15 to 20 bushels.

7. More efficient methods to control pests and diseases of both plants and animals will be available. The effectiveness of the new materials and improved techniques for applications will become more evident within the next few years.

8. Results from animal-breeding experiments will gradually increase the efficiency of livestock production. Work now under way is likely to produce hogs that are more efficient converters of feed into pork of the more desirable cuts. Dairy-herd improvement will be accelerated by more widespread use of artificial insemination.

9. Further improvements will be made in feeding methods. More adequate and better balanced rations will contribute to increased output per animal.

10. Some new land will be brought under cultivation by irrigation, drainage, and clearing, but the total new farm acreage is not likely to be large. If public development work now under way is continued and if all authorized work is carried out, about 4,500,000 acres will be brought under irrigation in the next 10 years. Around 8,000,000 acres might be improved by drainage or clearing during the same period. About half of these developments would take place on existing farms and the rest would involve bringing new areas into production.

11. Supplementary irrigation in humid areas has developed rapidly during the last few years. It is likely to be extended further, especially if the market demand for the products that are irrigated is sustained at fairly high levels.

12. If opportunities for employment are freely open in the cities, many small and unproductive farms will shift from full-time to part-time operation, or even become rural homes where little or no farming is done. If depression conditions should prevail for any length of time this movement might be reversed, as many unemployed people are likely to try to make at least a part of their living from the land.

13. As good roads, electricity, and other conveniences, become more readily available in rural areas more and more people engaged in nonfarm work will seek to establish rural homes. Thus the number of part-time farms and rural homes will be augmented from two sources: (1) Farm people shifting from full-time to part-time farming and (2) urban people seeking homes on the land.

14. Fewer workers will be needed in full-time farming as mechanization gains momentum in cotton production, and in other enterprises that now require much hand labor.

15. Family farms are likely to become larger and somewhat fewer as the productive capacity of farm workers is increased by the newer techniques. Some increase in the number of large-scale farms should be expected. They are not likely to constitute more than a small percentage of the total number of farms but they may produce a rather large percentage of total output.

16. Commercial farming will become a more complex business as technological advances continue. As family farms grow larger more capital will be needed for equipment and livestock. This means that adequate training and managerial ability of a high order will be needed for successful operation of commercial farms.

Changes will occur that are now unforeseen. For example, we have no way of foretelling the impact on agriculture of developments in regard to atomic energy. Over the next quarter-century innovations may be even more significant than those that are now on the horizon. But they are likely to be less important in the next few years because a period of testing of new developments is usually required, and later adoption by farmers is a gradual process.

EFFECT OF PROSPECTIVE CHANGES ON FARM OUTPUT

FARM OUTPUT LIKELY TO CONTINUE AT HIGH LEVELS

Prospective changes that have been outlined are preponderantly those that will tend to push the output of farm products higher and higher, instead of allowing them to recede toward the prewar levels. In their study of peacetime production adjustments, State committees estimated that under favorable economic conditions after the war it would pay farmers to produce at a level about 43 percent above the prewar average.[7] These estimates were based on average weather conditions. They gave consideration to maintenance of soil resources, and included the effects of the adoption of known improvements that would be profitable under conditions of prosperity. Farm output in 1948 was 40 percent above 1935-39, but growing conditions were unusually favorable in that year.

If economic adversity should prevail, the rate of increase in output would be slowed down, but even under unfavorable price conditions, it does not seem likely that the total output would be reduced substantially, unless weather were less favorable than the average. Any reduction that would come from the use of less fertilizer, or from attempts to reduce other variable costs, would probably be partly offset by the effects of the landward pressure of unemployed people.

Severe drought, or other unfavorable growing conditions, could reduce the level of output considerably. In a drought year, like 1934 or 1936, the output might drop about 20 percent. But it would increase again when growing conditions improved. Some of the improved practices—such as using hybrid seed corn and drought-resistant varieties of wheat and grain sorghum, summer fallowing, and contour farming—provide considerable protection against unfavorable weather. But on the other hand, the yield-increasing effects of fertilizer and some other practices would not be realized in case of severe drought. Crop loss from unfavorable weather is one of the major hazards in present-day commercial farming.

[7] See Footnote 3, p. 25.

Aside from this hazard, most of the changes that have already taken place, as well as those in prospect, seem to point irreversibly in the direction of increased production. When the transition to peacetime market outlets has been completed, food production at high levels may have to face market difficulties, unless high employment and purchasing power are maintained, and the channels of international trade are kept open. Financial and trade barriers may limit exports that would supply unmet food needs in other countries.

But regardless of the market outlook, there is no road back from the agricultural revolution that we have experienced. Attention therefore necessarily centers on mobilization for efficient and profitable peacetime agriculture instead of reconversion to a prewar situation that will never return.

If the belief still lingers that production will recede to prewar levels, under average weather, the steps that would be retraced should be considered. Farmers generally cannot go back to animal power because there are too few horses and mules now on farms. The annual colt crops do not begin to maintain the numbers. The mechanical-power phase of mechanization is here to stay, and it is the cornerstone of high-volume output for the market. Going back to open-pollinated corn, or to low-yielding strains of other crops would be decidedly unprofitable even in a depression. And more effective control of insect pests and diseases is likely to be continued, somewhat regardless of price conditions.

It is possible that less fertilizer and lime would be used in a depression, of course, although it would be poor economy in the long run to reduce yields in this way. It would be contrary to the national interest to fail to apply the fertilizer that is necessary for maintaining stable, soil-saving crop rotations. Similarly, temporary reductions in expenses could be made by not carrying out certain conservation practices, but these savings would be made at the expense of future productivity.

POTENTIAL CHECKS ON FARM OUTPUT

The only effective steps that can be taken to reduce the *total volume of output* are those that shift either capital, or land, or labor resources out of agriculture, or perhaps shift all three. But the preceding discussion indicates that if farming is to be carried on at all, an adequate supply of capital is needed for equipment and for current operations in order to achieve the most effective combination with the land and labor resources that are used in farming at any given time. In fact, many individual farmers have a tendency to invest too little capital with their land and labor resources; and changes in types of farming require new capital investment. This means that the primary steps in reducing the total volume of farm output would have to be taken by shifting either land or labor to other uses.

It is possible of course to use the agricultural land and the labor resources less intensively and to reduce output in that way. Land can be shifted from intertilled crops to sod crops of lower output per acre. And some lands now in crops can be returned to grazing. But farmers are likely to resist these changes for the reasons noted later.

Farm work days could be shortened. This would mean that less work would be done per worker. Farmers already have slackened the pace they maintained in wartime and it is highly desirable that many farmers reduce it even further. But more workers are now available on farms, and besides, with further mechanization, the farming can be done without working so hard. Consequently the effect on production of a shift to shorter working hours may be more than offset by more workers and more mechanization. In fact, the substitution of mechanical power for animal power increases the labor effectiveness of the farm family so much that usually there is soon the question of renting or buying additional land. If more land cannot be obtained an attempt is often made to farm the land more intensively. For example, soybeans are substituted for oats, dairy cattle for beef cattle, and the side-line poultry flock is expanded to an important enterprise in the farm business.

This tendency toward greater production *per farm* as a result of mechanization usually operates in the direction of more intensive rather than less intensive use of land resources. Therefore, if land is to be shifted from intertilled crops to sod crops, or from cropland to grazing, farmers must be convinced of the profitableness to them of this shift over a period of years. Otherwise public compensation is necessary to bring about such adjustments. Public programs could be developed that would shift some land to less intensive uses, but the effect on total output probably would be at least partly offset by more intensive use of other lands.

Some of the unproductive lands could be shifted out of arable farming by Government purchase or lease. They could be devoted to grazing or to forestry and recreational uses. Programs of this kind are needed in "fringe areas" of land that are poorly suited for agricultural production. But the total volume of output would be affected only slightly by carrying out such a program in poor-land areas. And public opinion probably would not support a program that would hold productive cropland out of arable farming for any considerable time.

It appears that about the only effective means by which total farm output could be reduced is by a shift of workers from farming to other occupations. That would occur only if employment off the farms were available for those who could not find attractive opportunities in agriculture. This question is discussed later, but it should be noted here that the shift would have to be large enough to result in a net decrease in the number of workers engaged in agriculture.

If a large number of workers shifted out of agriculture, however, the *per capita output* would rise for those who remained in farming. This in turn would mean higher *per capita incomes* with the same total output, and the economic reason for reducing output would disappear. But migration of this magnitude would be unlikely under conditions that result in low prices and pressure to reduce the total output of farm products.

The conclusion seems inescapable that no forces are now operating, or are likely to appear, that are sufficiently strong to offset much of the effect of the forces that will push agriculture in the direction of high-level production. Individual crop or livestock enterprises could be reduced in volume, either by voluntary shifting to other products or by production-control programs; but other products would be substituted and so total output would not necessarily be affected. To make substantial reductions in the total volume of farm output without shifting land and labor resources to nonfarm uses would require rigid controls. It would mean onerous restrictions on the use of land and labor in farm production.

OUTPUT IN RELATION TO POTENTIAL MARKETS

PRODUCTION MAY INCREASE AT SLOWER RATE

If the output of farm products for human use is likely to remain high, it becomes necessary to examine the prospective volume in relation to potential markets. As a basis for striking a balance between potential output and potential markets it seems desirable to summarize the main forces back of increased production into two groups. One is the shift from animal to mechanical power. The other is higher production per acre and per animal. New land development would be a third group, but this is likely to be a minor rather than a major factor in increased production. Irrigation of land now in arable farming would come under the heading of increased production per acre.

It appears that the shift to mechanical power will proceed at a rapid pace for several years. But its effect on total output will diminish progressively as numbers of horses and mules decline toward minimum levels. This means that increased output will then come mainly from higher production per acre and per animal. The rate of

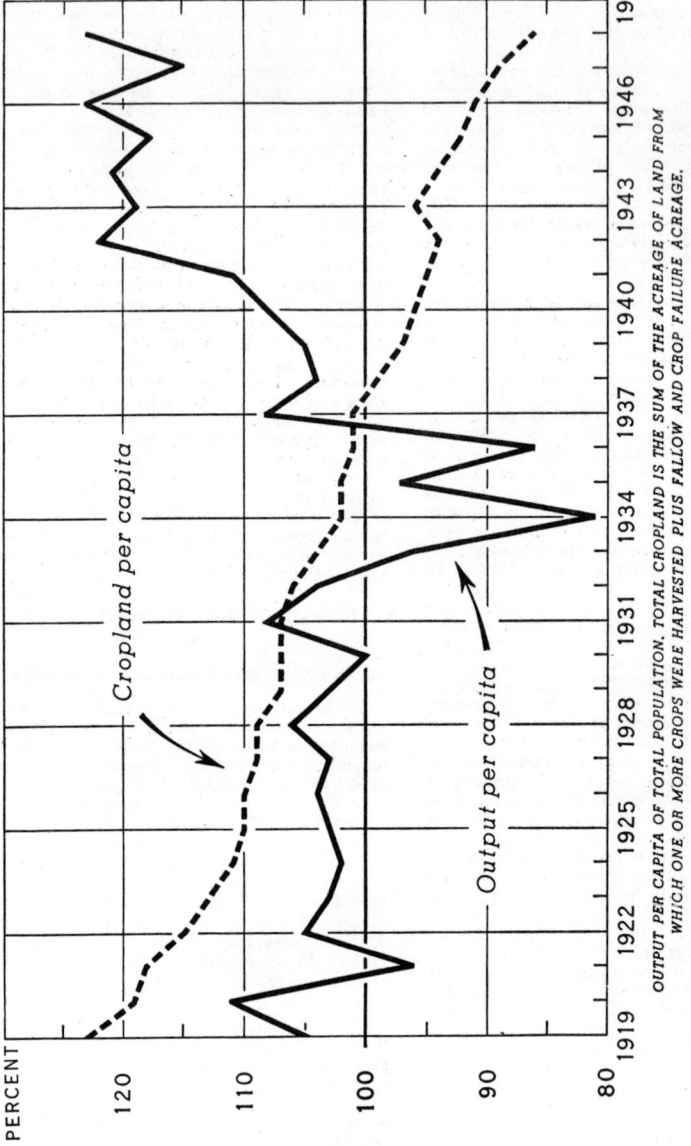

FIGURE 36.—FARM OUTPUT PER CAPITA, AND TOTAL CROPLAND PER CAPITA, UNITED STATES, 1919-48. (INDEX NUMBERS 1935-39=100.)

Cropland acreage per capita has decreased almost steadily for more than a quarter century, while the per capita output of farm products was maintained in most of the interwar years and increased sharply during the war. Release of cropland from production of horse and mule feed largely made possible the maintenance of output per capita in the interwar years. Increased crop production per acre was mostly responsible for the wartime expansion in output per capita.

such expansion depends upon new advances in technology, and their adoption by farmers.

It seems probable therefore that the rate of increase in production might slow down after the substitution of mechanical power has spent its force, unless new accelerating forces are introduced. The rate of increase for the next few years is not likely to be at the wartime pace, because part of that expansion represented release of production capacity that had been dammed up by drought and depression. But previous discussion indicates that an upward trend should be expected.

FORCES TENDING TOWARD LARGER MARKETS

On the market side the largest item is the level of domestic consumption of farm products. This depends primarily on the size of the population, and on the purchasing power that is available for consumers to buy farm products. The population of Continental United States is still increasing and with an upward surge in recent years. One is likely to forget that there were 27,000,000 more people in this country on January 1, 1949, than there were 20 years ago. By 1955, the population seems likely to be about 8,000,000 to 10,000,000 larger than in 1949.

When prospective levels of production are compared with some prewar period it is necessary therefore to remember that there will be many more consumers of meat, milk, and other farm products. A per capita comparison is much more appropriate. Figure 36, page 64, shows farm output per capita and cropland per capita by years from 1919 to 1948. Although the cropland per capita declined about one-fourth in that period the level of farm output per capita increased by 10–15 percent. Assuming average weather, the level of output per capita probably would now average about 10 percent higher than in the years 1925–29.

If exports and imports were maintained at the same levels as in the late 1920's, and if output increased only at the same rate as the population increased, the per capita domestic consumption of farm products would need to be only about 10 percent above the level of 1925–29 to absorb the total farm output. Average food consumption in the 4 years 1945–48 has been nearly 15 percent above that of 1925–29. Consequently it seems possible to achieve a level of domestic food consumption that would provide a market for a balanced output of farm products if high employment and purchasing power can be maintained, and if prices of the foods of higher value are in balance with consumer purchasing power. It should also be feasible to maintain the use of nonfarm food products at higher levels. In the 1920's a large segment of the population did not have the money to buy food enough for minimum adequacy. Consumption of nonfood farm products also was restricted by the low incomes of some groups. Moreover, the export market, although greater than in the 1930's, was not so large an outlet for farm products as it would be possible to develop in an expanding peacetime world economy.

This potential balancing of production and market outlets over a period of years is dependent upon three factors: (1) Increase of farm output no greater than the increase in population, (2) maintenance of employment and purchasing power that will support a per capita level of consumption of food and other farm products at least 10 percent above that of 1925–29, and (3) maintenance of at least as high a volume of exports as in the late 1920's. Any one of these three forces could upset the balance. If a spurt in technological advances should result in a much greater increase in output than in population farm products would press more heavily on market outlets. And if such a spurt should coincide with a period of unemployment the lowering effects on farm prices would be accentuated. On the other hand, maintenance of a high level of employment probably would result in a level of per capita consumption more than 10 percent above the 1920's. Some adjustment is to be expected with the subsidence of the world food emergency, but if other sectors of the economy are prosperous the markets are likely to absorb a large volume of farm products.

NEED FOR BALANCED PRODUCTION OF SPECIFIC PRODUCTS

Although under conditions of prosperity this large volume might be marketed without heavy downward pressure on farm prices, production of the different commodities would need to be balanced in relation to their respective market outlets. But that is where we are likely to have difficulties. Cotton, wheat, potatoes, and eggs, are already experiencing them. Oil crops are a potential trouble spot. On the other hand, more milk, meat, poultry products, fruits and vegetables, would be needed for high level nutrition in our country. And a shift from intertilled crops toward more hay and pasture, which in turn means more livestock, would help to maintain and improve soil resources.

If such shifts could be accomplished we would tend to utilize more effectively the Nation's resources in both land and labor. Beef production is a land-consuming enterprise, and milk, fruits, and vegetables are labor-consuming enterprises. But changes in these directions depend upon a farm price-and-income structure that would make them profitable to farmers. Under conditions of prosperity these adjustments could be made profitable because consumers would have the purchasing power to buy high-priced foods. But if depression conditions should prevail it would be more difficult to shift production in this way.

ADJUSTING MARKET OUTLETS TO BALANCED FARM OUTPUT

This rough balancing of potential production with potential market outlets indicates that the markets could absorb the potential production if conditions were favorable for high-level consumption. But work in market development would have to be undertaken at home. Export channels would need to be kept open. Shifts in production would be needed—in the direction of foods that would be emphasized in a prosperity diet, and toward farm products that would find larger industrial uses, or that could be exported in greater volume.

As markets for farm products can be expanded, and as the total volume of farm output cannot be reduced without severe restrictions on the use of land and labor, it appears that the slogan "Adjust production to market demands" of the interwar years might well be reversed. It was applicable to shifts from commodities that were pressing heavily on market outlets and into the production of others that had a brighter market outlook. But when reduced purchasing power limited the outlets for *all farm commodities* the remedy was not to be found in reduced production. If farm workers could have found other employment the balance might have been restored by reducing the *total output* of farm products, but the contracted purchasing power coincided with unemployment in the cities. And this caused a landward rather than a cityward movement of workers.

The experience of the 1930's suggests that neither the supply of, nor the demand for, individual farm products is as fixed and immutable as the laws of nature. Each can be modified by human effort, and by man-made institutions. But that experience also suggests that it is less difficult in times of depression to increase the market outlet than it is to reduce the total volume of farm output. From the standpoint of national welfare, food is needed just as much during depression as during prosperity. In fact, measures are needed to increase food consumption among certain groups of the population, even in prosperous times.

Human distress would be accentuated if farmers were to attempt to reduce the total volume of food output in depression years. But farmers do need some form of income insurance, or of income floor, that will protect them against disastrously low prices. Such protection is the farmers' counterpart of minimum wages and unemployment insurance.

Procedures can be developed that will promote a high demand for farm products—for domestic food consumption, for industrial uses, and for export. A national goal of adequate food for health available to all citizens, regardless of the state of economic activity, would go far in the direction of providing stable outlets for farm products. But this requires that public measures be devised to provide adequate food for those who cannot afford it, and that such programs be expanded as needed, if depression should come. Education in the elements of good nutrition could be greatly strengthened for all age groups and all income classes, and this also would help in achieving the goal of adequate food for all.

New uses may greatly enlarge the markets for some farm products in the course of several years. Research in this field holds considerable promise. But it must be clearly recognized that it is not enough merely to discover the physical suitability of a product for the new uses. It must be possible to produce this commodity at a cost that enables it to be sold in competition with alternative products. Price policies for specific farm products can either promote or retard the development of new market outlets.

International arrangements that permit a large volume of both exports and imports can supply a part of the outlet for farm products. Export markets for cotton, tobacco, wheat, and fruits, are extremely significant to all farmers, because competition in other products will be intensified if these outlets are not available. Perhaps even more significant, however, is the indirect effect of international trade in nonfarm products. Exports of automobiles and farm machinery, for example, create domestic employment and therefore increase the purchasing power for food and fiber.

A large volume of exports requires the acceptance of goods and services in exchange. Programs that would protect specific farm products at the sacrifice of greatly expanded world trade would create added competition in the domestic market on the part of farmers whose export markets had been destroyed by the trade restrictions.

Emphasis can be shifted in the direction of reducing costs or of developing more profitable alternatives for the products that might be injured by freer trade. If by mechanization and higher yields the cost of producing oil crops can be reduced materially, for example, they can be grown profitably and in large volume in competition with imported oils. If cost-reducing measures are not sufficient to achieve this possibility, more profitable alternative enterprises should be developed to replace the higher-cost part of the production of domestic oil. Farmers can then shift to those alternatives. Such measures would enable farmers to produce on an efficient low-cost basis in competition with other areas.

PROFITABLE AND ABUNDANT FARM PRODUCTION

Farmers want to produce both abundantly and profitably. They want to make full economic use of their resources. The nature of farm costs makes abundant production the most profitable use of their resources by individual farmers. But the resulting large output under certain circumstances might lower the prices of some products so much that production would become unprofitable for the entire group of producers.

COSTS IN RELATION TO PRODUCTION AND PRICES

Individual wheat producers, for example, cannot reduce output without lowering their individual net incomes, but the price-depressing effect of a large crop may reduce the income of all wheat farmers. The case is usually not so clear-cut as this, but it is natural to reach the conclusion that wheat farmers would increase their incomes if they all reduced production by some pro rata amount. At this point, however, the wheat farmer's cost structure should be examined. If through mechanization he already has labor and equipment on his farm that is partly unutilized, a restriction of output means that his overhead costs will have to be carried by a smaller quantity of wheat. Therefore, his average costs per unit of product will increase if production is reduced. Unless the land taken out of wheat can be shifted to some other productive use the net income available for the farm family might actually be lowered with a reduction in the acreage of wheat.

This illustration from wheat production indicates the need for analyzing costs in relation to production and prices, especially with reference to the effect of the changes in farming that are the primary concern of this report.

Improvements in farm technology frequently are associated with larger output of the products affected by the change. This is not always the case because some improvements save labor or capital investment without increasing the output, but most of them do result in more products. When the demand for farm products is expanding, at least as rapidly as the products going to market are increased by technological advancement, the market will absorb the larger output without reduction in prices. Farmers will benefit from improvements adopted under those conditions. The general economy also benefits because the larger output prevents a rise in prices. This was the case during the war and the early postwar years. But if production increases faster than the demand for the product, prices are likely to go down.

When that happens a part or perhaps all of the gain from the improvement may be shifted away from the farmers; and, for the general economy, may be offset by greater unemployment.

Whether an improvement lowers costs without affecting output or results in an increase of farm products, the farmers who first adopt it will retain whatever gain results, *until or unless prices of farm products are affected*. This means, of course, that farmers who adopt an improvement that actually reduces costs always gain in the early period of its adoption. The farmers who do not make the change are not affected by the improvement until or unless prices of farm products are reduced. But if improvement results in the displacement of labor, hired labor may be adversely affected, unless other employment is available that pays as well or better than the work from which they were displaced.

Farmers will tend to hold all of the gains from improvements that do not result in a larger output, because these changes have no adverse effects on prices.[8] The ultimate effects on farmers of improvements that increase production are not so clear.

In view of the emerging market difficulties in some farm products, it may be helpful to trace the economic effects of an improvement that results in a larger output. Hybrid seed corn is again a good illustration. Yields per acre are increased about 20 percent and the extra cost of hybrid seed is small in relation to this increase. For purposes of this illustration, we might take a 50-bushel yield with open-pollinated corn, and say that with the use of hybrid seed the yield was increased to 60 bushels, or 10 bushels per acre. For simplicity, let's assume that the price of corn is $1 a bushel. Then the additional income per acre is $10. Subtracting the higher cost of hybrid seed and of harvesting the larger crop may leave about $7 per acre net gain from the use of hybrid seed. This is an improvement that is easy to adopt and very profitable to farmers who make the change. Experience in the Midwest indicates that adoption is therefore rapid once the possibilities are known and the adapted seed is available.

As a result of widespread adoption of hybrid seed, the quantity of corn going to market might increase 10 percent, and if there is no offsetting increase in the market demand for corn the price of corn might go down 15 percent. The before-and-after situation of an average farmer with 50 acres of corn then might be about as follows:

---

[8] They do not affect prices unless they result in making farming so attractive that more labor and capital are invested in farm production. This could result in so bidding up the price of land that the gain would become capitalized, and new purchasers would not benefit because they would have a higher cost structure.

Before the change:
    50 acres at 50 bushels per acre, or 2,500 bushels corn.
    2,500 bushels at $1 per bushel, or $2,500 gross return.
After widespread adoption has resulted in lower prices:
    50 acres at 60 bushels per acre, or 3,000 bushels corn.
    3,000 bushels at $0.85 per bushel, or $2,550 gross return.

Although the assumed reduction in price shows a slight gain in gross return with the higher yield, the extra cost for seed and for handling a larger crop might actually mean a lower *net return* to the farmer for the larger crop.

If the price goes down so much that the larger output brings no more income to the farmer than he got before the improvement was made, the only way that he could continue to gain from the improvement would be to reduce his *total costs*. If this could be achieved he would be producing the larger output at a lower total cost than was formerly incurred to produce the smaller output.

It may seem difficult to produce more products at a lower total cost than was previously incurred for a smaller output but this has actually occurred rather generally on farms in this country over the last quarter century. For example, changes in farm power and machinery from 1920 to 1940 resulted in an actual decrease in both the investment in power and machinery and the current operating costs when the same prices are used in both periods. But this is not all. We have already seen that such equipment enables a man to do more work than he could with horses or mules and the old type of machinery.

Suppose that adoption of mechanical power on the same Corn Belt farm that adopted hybrid seed corn enabled the farm family to do the work with little or no hired help. They would then save both on the cost of power and on outlay for hired labor. The result would be a larger total output at a lower total cost, which would be accomplished by adopting a combination of improved practices. The combination frequently is extended to include the use of commercial fertilizer and more legumes in the rotation, which in turn means higher yields of corn and other crops. This chain type of reaction also includes improvements in livestock practices on many farms.

Frequently the process is worked out a little differently. Suppose the operator of the Corn Belt farm in our illustration decides to rent an extra quarter-section of land—one that was formerly operated by another family. This enlargement of the farm increases the *output per worker* very considerably. But we should note, of course, that another family is released for other types of employment. Fewer people are now engaged in farming, but the *total cost* of producing farm products is usually reduced by this kind of change. The cost reduction, however, is not in proportion to labor displaced because in part it involves a substitution of capital for labor. But usually the net income is increased for those who remain in farming. Whether the effects on those who leave the farm, and on the general economy, are favorable or unfavorable depends upon whether other employment is available for those who are displaced.

EFFECTS OF TECHNOLOGICAL CHANGES ON COSTS AND RETURNS

These illustrations point to the following effects on costs and returns from farming, and on farm people, of increases and decreases in production on farms, especially those caused by technological changes.

1. Because the demand for many farm products is such that at any one time a smaller output has a higher gross value than a somewhat larger output, an improvement that increases production may result in lower prices for the product. If demand is not increasing prices might be reduced sufficiently to cause farmers to lose all or nearly all of the gain from a cost-reducing improvement that is associated with a larger output.

2. But if, as a result of mechanization and other improvements *that are already adopted*, a farmer has equipment and family labor that is only partly utilized, a reduction in his *total output* will increase his average costs per unit of product. If production is reduced under those conditions it will be made in the part of his output that is produced at the lowest cost per unit. The price of the product, therefore, would have to rise considerably to offset the loss in income from cutting back on the part of the output that had the lower cost. Restriction of a single product such as wheat would have the same effect, unless the farmer could substitute some other product on the land taken out of wheat.

3. Many farmers have been able to adopt improvement combinations that have resulted in lower total costs for a larger output than they formerly had for a smaller output. Such changes enable them to hold much of the gain from improved methods, even if prices go down because of larger marketings.

4. If improvements that reduce costs result in fewer workers on farms the *net returns per worker* engaged in agriculture can rise even though the total gross income to agriculture is reduced. In other words, improvements that result in a larger output per worker also are likely to mean higher net returns per farm worker. This result can be expected if workers who are no longer needed in agriculture can shift out of farm work and into other employment.

5. It makes a big difference whether an improvement that displaces many hired laborers, as for example, the cotton picker, is introduced at a time when other employment is available for the

displaced laborers; or whether it comes in a period of unemployment when the workers who are displaced cannot find other work. If the displaced workers have to be supported by various kinds of unemployment insurance or work programs, the gain from the improvement in the way of reducing costs may be offset temporarily by the cost of work relief. A reduction in total output may have the same labor-displacing effect as a labor-saving improvement.

6. Farmers who are not in a position to carry out cost-reducing improvements will not be injured by other farmers adopting them *until or unless* the effect is felt in the form of lower market prices. But there will be a greater spread in net returns between those who adopt lower cost methods and those who do not adopt them. The effect of this is seen in the relatively low incomes per worker in the farming areas that have been least affected by the recent changes in farming.

DESIRABLE SHIFTS COULD BE PROFITABLE

If at the same time that farmers are reducing their costs the market outlets for farm products are expanded—because of increases in population, by maintenance of a high level of production and purchasing power in the nonfarm sectors of the economy, and by other means—farmers need not fear the tendency of production-increasing improvements to result in lower prices. In fact, they will benefit from both the larger volume of production and the lower cost per unit.

But even under those conditions, production within agriculture would still need to be balanced in relation to the needs for different kinds of farm products. The harmonious relationship between conservation and high-level nutrition has been mentioned. But shifts in the direction of more hay and pasture, that induce the production of more milk and meat, need to become *the most profitable production alternatives* in the areas where such shifts are desirable, if the changes are to be carried out by farmers.

If the market outlook indicates that there should be less emphasis on wheat, cotton, and some other cash crops, and more emphasis on hay and pasture for livestock feed; but if market forces are not sufficiently strong to bring about these shifts they could be accelerated in two ways. One would be to provide high support prices for milk, meat, fruits, and vegetables. This would encourage shifts in production toward such products. But these are commodities that consumers buy in much larger quantities when their prices are relatively low in relation to consumer purchasing power, so higher prices would cut off the part of the market that would be essential in achieving the goal of high-level nutrition. And the potential shift in this direction would then be severely limited by the smaller market outlet.

The other way to aid farmers in achieving both conservation and desirable shifts in production would be to assist in lowering the cost of producing the products that promote conservation and good nutrition. To achieve this, emphasis might be placed on improvements that increase the efficiency of producing these products. Aid might be extended in obtaining lime, fertilizer, and legume and grass seeds, and other materials that are needed in working out long-time farm plans for stable, soil-improving farming systems. Establishment of more stable tenure systems would give farm operators greater financial interest in soil maintenance and improvement. Many farmers would need educational assistance and management guidance also in carrying out such a program. A *combination of these measures*, as needed on individual farms, would go far toward making the desirable shifts in production the most profitable ones for farmers to carry out.

Assistance to farmers in adoption of cost-reducing measures that are specially applicable to hay, pasture, and livestock, should result in an expansion of the total market for farm products. Larger quantities of meat and milk will be bought if these products can be produced profitably by farmers at prices that are relatively low in relation to consumer purchasing power. Market expansion is more difficult in some of the other farm products. As shifts in this direction also will tend to conserve soil resources they are especially desirable at a time when such products as wheat and cotton are likely to have market difficulties.

INCREASING OUTPUT AND INCOME PER WORKER

Because increases in output per worker usually result in higher net returns per worker special measures might be developed in some areas to capitalize on the potentialities of increasing output per man as mechanization and other improved practices are adopted. Table 11 shows the progress in reducing hours of labor on corn, wheat, cotton, and some other crops, from 1910 to 1948. These changes indicate potentialities in reducing costs by lowering the requirements for labor in the major farm crops.

TABLE 11.—*Average hours of labor used per acre and per unit of production, and yield per acre for designated crops, selected periods 1910–48* [1]

Crop and item	1910–14	1925–29	1935–39	1940–44	1945–48
Corn:					
Man-hours per acre	35	30	28	26	24
Yield, bushels	26	26.4	25	32	35.2
Man-hours per 100 bushels	135	114	112	82	67
Oats:					
Man-hours per acre	16	12	10	9	8
Yield, bushels	29.4	29.5	29.2	31.8	35
Man-hours per 100 bushels	53	40	35	29	23
Hay:					
Man-hours per acre	12	12	11	12	12
Yield, tons	1.15	1.22	1.24	1.35	1.37
Man-hours per ton	10	10	9	9	9
Wheat:					
Man-hours per acre	15	11	9	7	6
Yield, bushels	14.4	14.1	13.2	17.1	17.7
Man-hours per 100 bushels	106	74	67	43	34
Rice:					
Man-hours per acre	55	37	32	29	26
Yield, bushels	35.8	42.9	49.7	45.5	46.4
Man-hours per 100 bushels	154	87	64	64	56
Potatoes:					
Man-hours per acre	76	73	70	71	80
Yield, bushels	99.7	114	117.2	136.7	182.3
Man-hours per 100 bushels	76	64	59	52	44
Sweetpotatoes:					
Man-hours per acre	132	122	116	115	118
Yield, bushels	94.4	93.8	84.9	87.4	96.3
Man-hours per 100 bushels	140	130	137	132	123
Dry beans:					
Man-hours per acre	47	30	28	24	21
Yield, pounds	778	655	855	898	988
Man-hours per 100 pounds	6	5	3	3	2
Sugar beets:					
Man-hours per acre	128	109	97	95	90
Yield, tons	10.6	10.9	11.6	12.7	13.2
Man-hours per ton	12	10	8	8	7
Cotton:					
Man-hours per acre	116	96	99	103	102
Yield, pounds	200.6	171.3	226.2	259.9	268.6
Man-hours per bale	277	268	210	190	182
Tobacco:					
Man-hours per acre	356	370	415	448	495
Yield, pounds	816	772	886	1,026	1,164
Man-hours per 100 pounds	44	48	47	44	43
Soybeans:					
Man-hours per acre		16	12	11	10
Yield, bushels		12.6	18.5	18.3	19
Man-hours per 100 bushels		126	64	58	52

[1] Hours of labor are computed for the acreage harvested and include preharvest work on acreage that was later abandoned.

During the war, the labor used per unit of product was reduced along the entire farm front. Gross farm production increased 18 percent from 1939 to 1944, but the total man-hours expended for the 1944 production were only slightly more than those used in 1939. And as there were fewer workers on farms, the production per worker in 1944 was 26 percent above the 1939 level. Farm production in 1944 used about 3 billion fewer man-hours than would have been needed under 1939 production conditions. That represents a saving of 1.5 million man-years of work at 2,000 hours per worker. The change is accounted for by (1) lower labor requirements per unit of product resulting from increased crop yields per acre, (2) greater mechanization, (3) some shifting out of high-labor crops such as cotton, (4) less labor per unit of livestock because of increased production, and (5) the spreading of overhead and miscellaneous labor over a larger volume of production.

Figure 37, page 72, summarizes the remarkable gains in efficiency that agriculture has made over the last quarter-century. Total physical inputs (farm labor, power, and other resources) per unit of farm products decreased about one-fourth from 1920 to 1945. This means that the cost per unit of producing farm products has decreased about 25 percent in terms of the physical labor, power, fertilizer, and other materials that go into farm production.

These accomplishments represent rapid progress, but much more startling changes lie ahead if full advantage is taken of the opportunities for reducing labor requirements and for increasing the output per man. For example, mechanical pickers and other improvements may soon make it possible to reduce the man-hours used per bale of cotton in many areas to about one-fourth of the hours required by the older methods. If fairly complete mechanization of the cultivation and harvesting of cotton could be attained, the man-hours used per bale might be reduced to 65 hours, as a national average. This would release 1.6 billion hours of labor on an output of 13,000,000 bales. That is equivalent to about 800,000 man-years of labor at 2,000 hours annually per worker. Expenses per bale of cotton would be greatly reduced. Such a drastic change could not come rapidly, but the end result would have a decided impact on the economy of the South, as well as on the entire national economy.

Changes that would enable agriculture to take full advantage of the potentialities of the new technology would be so sweeping that there is little likelihood that they will ever be completely realized. In many respects this may be fortunate, because national welfare cannot be measured solely in terms of efficiency. But it seems apparent that considerable progress in adapting farming to the new technology is desirable and necessary if farmers as a group are to be prosperous.

The distance to be traveled varies greatly by areas and by broad regions. Figure 38, page 73, shows the gross farm production per worker by census geographic divisions from 1919 to 1948, compared with the average production per worker in the United States for the years 1935–39 as a base, or 100 percent. The effect of the drought in the early 1930's and the piling up of workers on farms because of the depression is evident in the curve that shows gross production per worker for the United States. But the influence of drought on production per worker is most evident in the north central and mountain divisions.

Gross production per worker in the three southern divisions has been much lower than the 1935–39 average for the United States throughout the entire period shown in figure 38, page 73. These three southern divisions have had about 50 percent of all the farm workers in the Nation, and have contributed about one-third of the total United States production in most of the years of the interwar and World War II periods. With the exception of the years of drought, gross production per worker in the West North Central States was 50 to 60 percent above the 1935–39 national average in the interwar period, and it was more than twice that prewar national average in the war years, 1942–44. Gross production per worker in the Pacific States has shown a tendency to increase throughout the entire period.

What are some of the causes of such wide geographic variations in gross production per worker? Table 12 gives part of the explanation. The South Atlantic and East South Central States had the lowest gross production per worker in 1944. The acres of cropland, value of land and buildings, and value of livestock per worker, were less than one-half of the national average. The value of equipment per worker was about one-third of the average for the country as a whole. In other words, the average farm worker in these two southern divisions had less than half as much land, buildings, and livestock, and had only about one-third as much machinery to help him in his farm-production job, as did the average farm worker in the United States.

The figures in table 12 indicate that material increases in production and net income for southern farm workers largely depend upon (1) providing more land, livestock, machinery, fertilizer, and other capital items per worker and (2) opportunities for nonfarm work for the young people who grow up on farms but who will not be needed in farm occupations, and for the workers who will be released from agriculture as mechanization and other improvements gain momentum. These changes are inevitable. They are already under way. The only question is how rapidly the transformation will take place.

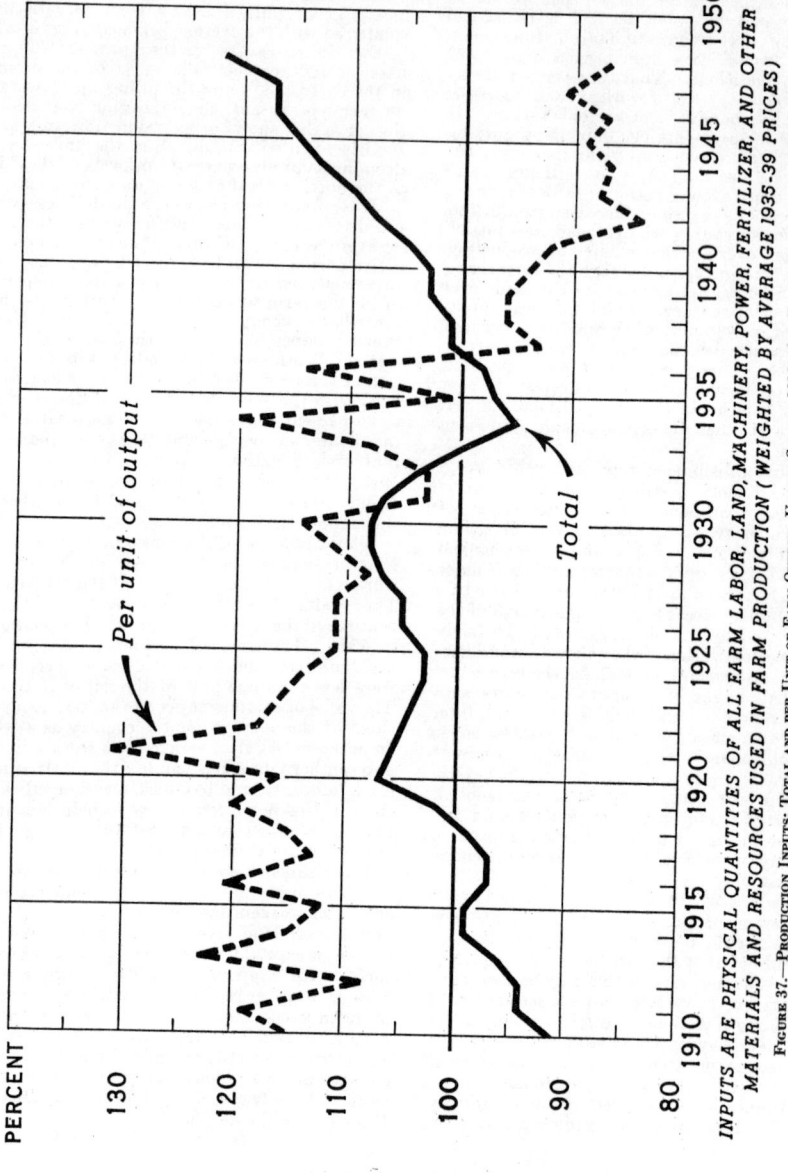

FIGURE 37.—PRODUCTION INPUTS: TOTAL AND PER UNIT OF FARM OUTPUT, UNITED STATES, 1910–48. (INDEX NUMBERS 1935–39=100.)

Total physical inputs (farm labor, power, and other resources) per unit of farm output decreased about 25 percent over the last quarter century.

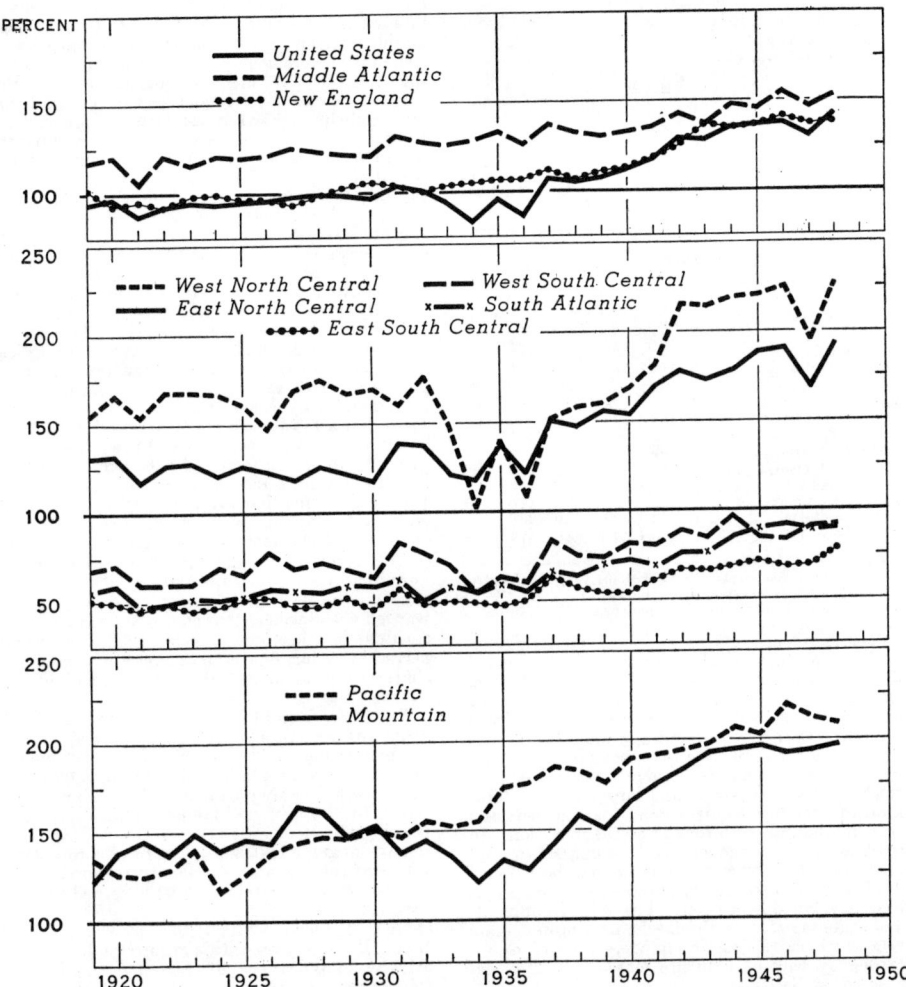

FIGURE 38.—RELATIVE GROSS FARM PRODUCTION PER WORKER, BY GEOGRAPHIC DIVISIONS, 1919–48. (INDEX NUMBERS U. S. 1935–39 AVERAGE=100.)

Farm production per worker has been lowest in the southern divisions during the entire period 1919–48. The most consistent upward trend in production per worker has taken place in the Pacific division, although relatively greater wartime increases occurred in the North Central and Mountain divisions.

TABLE 12.—*Gross production per worker, and value of land and buildings, livestock and equipment per worker, by census geographic division, 1944 and 1945* [1]

Census geographic division	Production per worker, 1944 (U.S. average=100)	Land and buildings per worker, 1945 [2]	Livestock per worker, 1945 [2]	Equipment per worker, 1945 [2]	Total cropland per worker, 1944
	Percent	Dollars	Dollars	Dollars	Acres
Pacific	154	8,746	826	623	32
Mountain	144	6,470	1,820	719	68
Western States	150	7,784	1,246	664	48
West North Central	162	7,668	1,592	926	91
East North Central	132	7,175	1,184	863	45
Middle Atlantic	109	3,942	1,018	838	24
New England	100	3,960	745	529	17
Northern States	139	6,665	1,299	864	59
West South Central	71	3,433	616	320	32
South Atlantic	63	2,212	343	187	15
East South Central	52	1,923	351	177	15
Southern States	62	2,519	434	227	20
United States	100	4,622	844	513	38

[1] Production per worker in 1944 is used because that is the production year reported in the 1945 census.
[2] From the 1945 Census of Agriculture.

Readjustments of this kind will create opportunities for farm workers in the South to equal the per capita production and income of farm workers in other regions. But on many farms the change involves shifting from a simple cash-crop type of farming, with hand-and-mule operations, to relatively complex soil-conserving types, that involve forage, pasture, winter cover crops, livestock, mechanical power, and larger farms. This kind of farming requires more management and more mechanical skill for successful operation. In most other areas the shift to mechanization and more complex types of farming has been made gradually, and managerial and mechanical skills have developed as needed. The shift from hand-and-mule farming to the mechanical-power phase is a much more drastic change. It may be retarded by lack of requisite managerial and mechanical skills.

Eventually southern farmers will learn the new ways of farming even by trial and error methods; but it would be possible to speed up the learning process. A *management advisory service* could help to overcome the lack of experience with new methods. Such a service could help farmers work out systems of farming that involve the improved methods, and it could also provide some guidance in adopting the new techniques. Encouragement might be given to the organization by farmers of cooperative-management associations for employing professional management assistance. Research agencies could render service on this front by establishing research test farms where new developments could be tested, and where farmers would see them in operation. Frequently there would also be need for credit programs to provide capital for improvements and for enlarging farms.

Although the readjustments in southern farming are likely to be more drastic than changes in other areas the same general process is at work elsewhere. Mechanization and other improvements have not only increased the size of farm that a family can operate but they have also made farming a much more complex business. This means that successful, modern farming requires less brawn and more brains than the farming of a generation ago, and that it takes a large capital investment to become established on the size of farm that a family can easily handle from a labor standpoint. Some farmers will need new sources of credit to obtain farms of adequate size, and to buy the livestock and equipment to operate them. Full-time farms that are too small to utilize mechanical power effectively are likely to be too small to provide a satisfactory living for a farm family.

The farmers who can utilize available assistance in adjusting the type and size of their operations to the new techniques are likely to be rewarded with incomes that compare favorably with earnings in other occupations. Such rewards are necessary if farming is to attract a proportionate share of the capable youth who are choosing their life occupations.

But tables 8 and 9 indicate that there are large groups of farm people who have not benefited from mechanization and associated improvements. Some of these people probably are best fitted for farm work, but they cannot readily adapt themselves to the new techniques. Many are older people who look forward to retirement. But some younger workers will not be able to operate farms of the size and complexity required by the new techniques. Greater effort is needed to assist these small-scale farmers in making the kind of changes that are most likely to improve their situation. Perhaps part of this group can develop relatively simple types of farming which, even though somewhat less profitable, are better suited to their capacities. Research is needed on the economic possibilities in this direction. Some will find their best opportunities as hired farm workers. Better housing and more adequate security are problems of major importance to the hired-labor groups.

The largest group among those who are likely to be disadvantaged by the further shift to the mechanical-power phase and associated improvements are the ones who will have to seek employment elsewhere because their work will no longer be needed in agriculture. If high industrial activity is maintained, other work will be available. But even so, the shift to new work and new environment will be difficult. Employment offices in rural areas will be needed to inform workers of job opportunities. Many of those who are no longer needed on the farms will need preliminary training for other occupations.

As modern transportation makes it possible for more people to combine rural living with nonfarm work, it would be possible to expand the number of part-time farms and rural homes at the same time that many workers shift to nonfarm employment for their major source of income.

A high level of economic activity will need to be maintained if the workers displaced in agriculture are to be absorbed in other occupations. Figure 6, page 9, indicates that progress in technology has increased output per man in other industries in the same way as in agriculture which means that a market must be found for a greatly increased industrial output. To a very large extent, increased employment at high wages can create its own market, if consumers are permitted to share the benefits of improvement in the form of lower prices. But as output from the productive plant of this country is increased it may become highly desirable to shift more effort into the service occupations—into the health and educational professions, into recreation, and into other professions and services that will provide better living for all the people.

A high level of production is not an end in itself. It is only one of the means to achievement of a better way of living. As increased efficiency is developed in both agriculture and industry less work will be required to produce both food and other products. More time will be available for other things including leisure and recreation.

Farmers can retain for themselves some of the real benefits from agricultural improvement if they will take effective steps to utilize the first results. These steps include (1) slackening the pace of farm work and increasing the leisure time available for the entire farm family, (2) investing the higher earnings in education and health, and in home conveniences, and (3) refraining from capitalizing increased earnings into higher land values, which make the business of farming more hazardous and less profitable for themselves and for their children. Better farming should always mean better living.

NOTE

This report is a revision of the summary of a study that was begun in the fall of 1944 with the purpose of analyzing the changes in farming during the interwar and war years, appraising the forces back of the large increases in production, and evaluating some of their peacetime implications. The original summary report was issued in processed form in June 1946 under the title "Changes in Farming in War and Peace." This is the second revision of that earlier summary but is the first report on the subject that has been printed. It includes our production experience in the United States in the years 1946 to 1948; and other new information, especially with respect to farm classification and trends in sizes and ownership of farms.

World Food Supply

An Arno Press Collection

Agricultural Production Team. **Report on India's Food Crisis & Steps to Meet It.** 1959

Agricultural Tribunal of Investigation. **Final Report.** Presented to Parliament by Command of His Majesty. 1924

Bennett, M. K. **The World's Food:** A Study of the Interrelations of World Populations, National Diets and Food Potentials. 1954

Bhattacharjee, J. P., editor. **Studies in Indian Agricultural Economics.** 1958

Brown, Lester R. **Increasing World Food Output:** Problems and Prospects. 1965

Brown, Lester R. **Man, Land & Food:** Looking Ahead at World Food Needs. 1963

Christensen, Raymond P. **Efficient Use of Food Resources in the United States.** Revised Edition. 1948

Crookes, William. **The Wheat Problem.** Revised Edition. 1900

Developments in American Farming. 1976

Dodd, George. **The Food of London.** 1856

Economics and Sociology Department, Iowa State College. **Wartime Farm and Food Policy,** Pamphlets 1-11. 1943/44/45

Edwards, Everett E., compiler and editor. **Jefferson and Agriculture:** A Sourcebook. 1943

Famine in India. 1976

Gray, L. C., et al. **Farm Ownership and Tenancy.** 1924

Hardin, Charles M. **Freedom in Agricultural Education.** 1955

High-Yielding Varieties of Grain. 1976

[India] Famine Inquiry Commission. **Report on Bengal.** 1945

Johnson, D. Gale. **Forward Prices for Agriculture.** With a New Introduction. 1947

King, Clyde L., editor. **The World's Food.** 1917

Marston, R[obert] B[right]. **War, Famine and our Food Supply.** 1897

Mosher, Arthur T. **Technical Co-operation in Latin-American Agriculture.** 1957

The Organization of Trade in Food Products: Three Early Food and Agriculture Organization Proposals. 1976

Projections of United States Agricultural Production and Demand. 1976

Rastyannikov, V. G. **Food For Developing Countries in Asia and North Africa:** A Socio-Economic Approach. Translated by George S. Watts. 1969

Reid, Margaret G. **Food For People.** 1943

Schultz, Theodore W., editor. **Food For the World.** 1945

Schultz, Theodore W. **Transforming Traditional Agriculture.** 1964

Three World Surveys by the Food and Agriculture Organization of the United Nations. 1976

U. S. Department of Agriculture, Agricultural Adjustment Administration. **Agricultural Adjustment:** A Report of Administration of the Agricultural Adjustment Act, May 1933 To February 1934. 1934

U. S. Department of Agriculture. **Yearbook of Agriculture, 1939:** Food and Life; Part 1: Human Nutrition. 1939

U. S. Department of Agriculture. **Yearbook of Agriculture, 1940:** Farmers in A Changing World. 1940

[U. S.] House of Representatives, Committee on Agriculture. **Oleomargarine.** 1949

[U. S.] National Resources Board. **Report of the Land Planning Committee. Part II.** 1934